D0368583

The Last Walk

THE

Reflections on Our Pets

LAST

at the End of Their Lives

WALK

Jessica Pierce

THE UNIVERSITY OF CHICAGO PRESS

CHICAGO AND LONDON

Jessica Pierce is a bioethicist and coauthor of
Wild Justice: The Moral Lives of Animals.

The University of Chicago Press, Chicago 60637
The University of Chicago Press, Ltd., London

21 20 19 18 17 16 15 14 13 12   1 2 3 4 5

ISBN-13: 978-0-226-66846-8 (cloth)
ISBN-13: 978-0-226-92204-1 (e-book)
ISBN-10: 0-226-66846-0 (cloth)
ISBN-10: 0-226-92204-9 (e-book)

Library of Congress Cataloging-in-Publication Data

Pierce, Jessica.
The last walk : reflections on our pets at the end of their
lives/Jessica Pierce.
pages ; cm
Includes bibliographical references and index.
ISBN-13: 978-0-226-66846-8 (cloth : alkaline paper)
ISBN-10: 0-226-66846-0 (cloth : alkaline paper)
ISBN-13: 978-0-226-92204-1 (e-book)
ISBN-10: 0-226-92204-9 (e-book)   1. Pets—Death—
Moral and ethical aspects.   2. Death—Moral and ethical
aspects.   3. Euthanasia of animals.   4. Euthanasia.
5. Human-animal relationships.   6. Dogs—Biography.
I. Title.
BD444.P54 2012
179'.3—dc23

2012007983

♾ This paper meets the requirements of ANSI/NISO
Z39.48–1992 (Permanence of Paper).

In honor and memory of Odysseus

Contents

1

Final Odyssey

Ody shuffles down the hall and stops at the doorway of my office, peering in at me with brown eyes made milky by age. He doesn't come all the way into the room to put a muzzle on my lap or push a nose under my hand as he used to. For Ody the greeting remains incomplete, a reminder that he now inhabits a different world.

I turn in my chair and call him. Though he doesn't come, I know he hears me. His stump of a tail flicks back and forth in reply. I also know, because we repeat this exchange day after day, what comes next. With a snort and a raspy cough, Ody will turn stiffly and make his way back down the hall, the click-drag click-drag of his nails telling me where he is headed. But I don't want him to go just yet.

I stand and step into the doorway. Kneeling, I take Ody's face in my hands. His long ears are like velour under my fingers. I run my hands along his body, feeling the spongy lumps that bulge out here and there, like a super-sized Braille inscription. The lumps, the vet tells me, are fatty deposits called lipomas and are a harmless, if unsightly, manifestation of age. Despite his lumps and skin tags and white hair, Ody is still just as handsome to me as ever.

Repeating another familiar exchange, I lower my face and touch my nose to his. I've always loved his nose, which is improbably colored to match the russet of his coat. I close my eyes and feel the cool roughness. His breath is a reminder of worn and broken teeth and of gums decayed by time. We remain here nose-to-nose for several long moments, and I then I stand up and turn back to my work. Ody shuffles off, click-drag, down the hall.

Ody is just over fourteen, and if you saw him on one of his occasional walks (he walks when the mood is right, and otherwise refuses to leave the house) you would know that he is an old dog. His back legs are atrophied and weak and bend awkwardly, and he stands as if he were halfway toward sitting. Every few steps one of his back legs fails to do its job, and he lands on top of his toes, rather than on his paw pad. Without support of the foot, the leg collapses, and his body dips and sways. This idiosyncrasy is most likely the result of some neurological dysfunction that causes the brain to send the wrong signals to the legs. It is one among several symptoms of "cognitive dysfunction syndrome"—in other words, Ody suffers from dementia.

Ody is nearing death. And the closer he draws toward the end, the more puzzled I become about what a good death would mean for him. It is pretty clear what a bad death looks like, and far too many animals in our world suffer a bad death, dying afraid, in pain, and alone or with strangers.

But what is a good death? The message I get from everything I read and all the people I talk to is that eventually Ody will reach a point at which his life becomes burdensome, and he will tell me, somehow, that he wants to be released. I will take him to the vet and the kindly people there will poke him with a needle and it will all be very quick and painless and gentle. But something about this scenario bothers me, like a splinter just under the skin of my conscience. And the closer Ody limps and shuffles toward this elusive endpoint, the less comfortable I become.

Is a "natural" death preferable, for Ody, to euthanasia? Why is it that we have such a revulsion against euthanasia for human beings, yet when it comes to animals this good death comes to feel almost obligatory? If it is an act of such compassion, shouldn't we be more willing to provide this assistance for our beloved human companions as well?

I worry: will I be able to read Ody's signals? And I wonder: does life ever become so burdensome for an animal that he or she would prefer death, or is this something we have judged from the outside? Is it that their lives become burdensome for them, or for us? The more troublesome Ody becomes—the more he pees on the floor, the more often he barks for no reason at odd hours of the night, the more frequently he stands, confused and panting, in the middle of the kitchen while I'm trying to cook dinner—the more ambiguous the question of burdens becomes.

WHY I WROTE THIS BOOK

When Ody was about thirteen and a half, I decided to keep a journal about his life. Although he was still in relatively good health, I could see age wearing its tracks onto his body and mind. His health was starting to fail in small ways—he had had mast cell tumors removed from his ear and from his haunch, his hearing was fading, and he had to work a little to stand up. I began to write down the funny and annoying things that Ody spent his days doing, so that I would remember him in color and detail. And I recorded my reactions to watching him grow old. I thought it might help me work through the anguish of someday losing him, and the difficult decisions that I suspected lay in wait for us. I didn't know it at the time, but the "Ody Journal" was the beginning of this book.

Ody's story soon became something more than personal. As a bioethicist, my work has focused on how the biomedical sciences intersect with human values, particularly within the context of healthcare. At the same time as I began writing my daily journal about Ody, I was finishing a large college-level textbook on bioethics. The ethics of death and dying has long been at the core of this field of applied philosophy, and one of the central chapters in the textbook focused on ethics at the end of life. I would sit at my desk, immersed in the literature about human death and dying and hear Ody retching in the background, as the water he just drank got stuck in his throat. I would have to get up from my work, frustratingly often, because he needed to go out and pee, again, or was barking at the door. It became obvious to me that many of the questions under discussion in human end-of-life care were similar to ones I might soon encounter with Ody. How aggressively should I treat his encroaching disabilities? How do I judge the quality of his daily life, as he experiences it? Might there ever come a time when his day-to-day living is filled with so much pain and fear that the humane course will be to hasten his death?

Bioethics has not generally concerned itself with animals, and most certainly not with the aging, dying, and death of animals. But as I began dealing with Ody's aging, and thinking about how to navigate decisions at the end of his life, I realized that end-of-life care for our animal companions is worthy of sustained attention and that pet owners and veterinarians face moral quandaries every bit as complicated as those we face with human loved ones. Yet as I learned the hard way, we are not always prepared for these challenges, not having been asked to think carefully through the terrain ahead of us. No one told me that having an old dog would be hard and that his approaching death would strike so much fear into my heart. I didn't know that planning for his death—knowing how a good death might best be accomplished, what might happen with his body, how I might find constructive ways to grieve—would have helped me do it all better, without so many regrets.

Soon enough, the bioethics book was done, and before me sat, clear as day, my next project: to write about caring for our aging and dying animals. I thought that I might, through research and deliberation, know what to do for Ody, when the time came. And I thought that hearing Ody's story might help others deal with the dying of a beloved animal companion.

I say this is Ody's story, but it is really my story, too. It is my story of watching an animal I love grow old, suffer the infirmities of age, and begin his descent toward death. It is my story of choosing and not choosing; of action and inaction; of coming to terms with change; of accepting the inevitable; and of holding his life in my hands and trying to figure out what to do with it. I began to worry about Ody's death long before he even began growing old. And it scared me. It was always crouching in the back of my mind, like an animal tracking its prey from the shadows: the fear that someday I would have to choose to "put him down," or watch him suffer; that I would have to play God. And that time did come, eventually.

But before we rush ahead, let's go back to the beginning.

MEET ODY

Ody joined our family as a ten-week-old wriggling sack of loose red skin. It was 1996. I was thirty. I had wanted a dog since forever, and I finally felt that I was in a position have one: settled into a life that rotated around the home. And once I wanted one, I wanted one badly. Some women in their thirties become obsessed with babies; I wanted puppies. But my husband Chris would take some convincing. I dropped hints here and there, but not too pushy. Not too needy. Subtle at first and then growing more blatant as he grew accustomed to the idea.

I didn't care what kind of dog—any would do. So when Chris mentioned that he might be interested in a Vizsla, without even asking the obvious "What's a Vizsla, and what are they like?" I immediately started scanning newspaper ads and kennel listings. Vizslas are not common, so I was readying myself for a hard search. But within only a few days, I opened the *Omaha World Herald* to the want ads and saw "Vizsla Puppies, $200." It was a Sign. By that afternoon we had met Ody and his aptly named brother,

Tank, the only two puppies left. We watched as the two pups wrestled and bit. One was clearly the boss—confident and way too busy to pay us any attention. The other, the smaller one, was sweet and friendly, and when he could push his way out from under Tank he would waddle over and crawl into our laps. We fell in love with Ody, and the rest, as they say, is history.

The Vizsla (pronounced VEESH-lə), also known as the Hungarian pointer, is considered a "general utility gundog." Ody, being Ody, is terrified of gun sounds and will start to pant and quiver if you so much as pop a plastic bag or squeeze too hard on bubble wrap. Vizslas are light and sinewy—the average male weighs about fifty pounds. Ody is a particularly stout Vizsla—not fat, but broad-chested and muscular, all seventy-five pounds of his short-haired rust-colored body. People who know Vizslas always remark on Ody's too-short tail. A perfect Vizsla would carry two-thirds of his tail; Ody has about one-third. Whoever did the tail docking must have slipped with the scissors. Still, it amazes me how much he can say with his stump: it embodies his personality and is a clear barometer of his mood. Pointed up (happy, excited), curved down (scared, upset), or pushed out straight as a rod behind him (squirrel mode). When Ody's stump wags—which it does all the time—his whole body shakes back and forth.

According to my Vizsla owner's manual, the breed originated with the Magyar hordes who during the tenth century invaded what is now Hungary. The dogs became prized hunting companions of the Hungarian aristocracy, who protected the purity of the breed through the centuries. During the Soviet occupation of Hungary after World War II, Vizslas were smuggled out of the country, and eventually they began arriving in the United States. We've always considered Ody an exceptionally regal dog—in looks, at least. And that explains the name Odysseus: a great king—handsome, cunning, beguiling—who undertakes an epic journey. Actually, his full legal name, if you must know, is Sadie's Rigorous Odysseus. His mother was Sadie, and his father was Rigor (with a brother named Mortis) and belonged to a mortician.

The Vizsla is said to be an expressive, loving, and gentle dog. Though intelligent and highly trainable, they are easily distracted and have a proclivity for stubbornness. The breed is highly tactile, and many Vizslas have an odd quirk: they like to hold people's hands in their mouth. Ody has never done this, but he does love to be touching someone whenever possible. He leans against you whenever you stroke his back, always sleeps in the bed under the covers, and fancies himself a lap dog. Vizslas are often referred to as "Velcro dogs," wanting to be close to their owners as much of the time as possible and with a tendency toward separation anxiety. Vizsla books and websites emphasize that these are extremely athletic dogs, requiring a great deal of exercise and stimulation. One source recommends that you run or walk your Vizsla at least six miles a day, preferably more. *That kind of mileage is outrageous for most people*, I think, though with a smug little smile and a self-righteous pat on my own back. Here, I think, is one thing I actually did right. For his whole life, Ody has been my running and mountain biking partner, and we covered a lot of miles in our day.

As I write this, Ody is asleep on the tan couch to the left of my desk. I feel a little stab of sadness every time I glance over at him, and more often than I'd like to admit, my eyes well up. The thought of losing him hurts, but even more than this, I mourn *his* losses. I am sad that Ody can no longer run wild through a field of tall grasses or chase the teasing squirrel in the backyard. But how do I know that he has lost anything, in his own mind? Why should I think that his diminished mobility makes him feel frustrated? How well do I really know Ody?

Despite the close bond we share, Ody is mysterious to me. The word that comes up over and over, when I think about Ody, is "inscrutable." And in his golden years, Ody has become, if anything, more difficult to read. In *Dog Years*, Mark Doty writes, "No dog has ever said a word, but that doesn't mean they live outside the world of speech. . . . To choose to live with a dog is to agree to a long process of interpretation—a mutual agreement, though the human being holds most of the cards." I try to move beyond the world of human language, into Ody's own form of speech, but rarely do I feel confident about my translation. In our relationship, I think Ody holds more cards than I do. Doty goes on to say, "Love for a wordless creature, once it takes hold, is an enchantment, and the enchanted speak, famously, in private mutterings, cryptic riddles, or gibberish." We have our private mutterings, Ody and I, but when I ask myself questions such as "Is Ody happy?" and "Is he suffering?" I find that I really do not know.

Wondering about Ody makes me wonder about animals in general. Are animals aware of their aging, of illness, of the dusky shadow of the grim reaper following behind them? What is their aging, dying, and death like *for them*? Very little sustained attention has been given to these questions. The presumption has long been that animals are not complex enough creatures and that dying and death are too abstract for any but the human mind to grasp. Even among those who fight for improved animal welfare, the focus of attention is almost always on the quality of animal *lives*. And this, of course, is paramount. But we mustn't neglect the quality of their death—particularly because it is we who often orchestrate their end. The ideal of a "good death" applies not only to human beings but also to our animal kin.

THINKING ABOUT ANIMAL DEATH

It is worth inquiring, first of all, how animals actually *do* die. It is impossible to say how many companion animals die each year in the United States since no one keeps a registry, as we do for human deaths. So this is merely educated guesswork. During one year, US consumers will have purchased or otherwise acquired an estimated 15 million birds, 94 million cats, 78 million dogs, 172 million freshwater fish, 14 million reptiles, and 16 million small animals. How many of these animals die each year and by what means is difficult to figure. Narrowing our attention to dogs and cats (since there are no data on other kinds of pets), cancer, kidney disease, and liver disease are the leading causes of death, in terms of disease processes. Yet the *main* cause of death in dogs and cats in the United States—that is to say, the central mechanism of death for most canines and felines—is undoubtedly euthanasia. The Humane Society of the United States estimates that six to eight million cats and dogs enter shelters each year, and three to four million are euthanized. No data are available for the numbers of dogs and cats euthanized each year in veterinary offices and homes; all we know for sure is that far more die by the needle than by natural causes.

What about the birds, fish, reptiles, and small animals? No one really has any idea how these pet animals die. There is no crime in buying a pet and killing it, as long as you don't do it on purpose or with cruel intent. My guess is that the vast majority of deaths occur through inadequate care. Animals simply wither away, perhaps because they don't have enough heat, or too much, or not enough moisture, or too much, or not the right kind of food. But "wither away" doesn't put quite a fine enough point on it, does it? These creatures—the corn snakes, hermit crabs, leopard geckos, and bearded dragons—die slow and unpleasant deaths after protracted, though perhaps unnoticed, suffering. I call this category of death lethal neglect.

Many animals die before they even have a chance to become somebody's pet: they die in pet stores, on the way to stores from breeders, and onsite at the breeders. Our local big box pet store has a shelf of Siamese fighting fish for sale—only $2.99!—each one in its own little tiny plastic cup. For some reason, I am drawn to this shelf of fish, and every time I enter the store I go look at them. And every time, at least two or three will be floating at the top of the cup. At this same store, company policy is that animals who escape from their cages in the store can no longer be sold, and guess what happens to them? We happen to know one of the managers, and she has over sixty pet rats at home, all rescued from the store because she couldn't bear to see them killed. A few years ago, she talked us into adopting an ugly hairless rat who couldn't be put out on the floor because he had been attacked by other rats and had bloody scabs all over his body. He had to wear a teeny rat-sized Elizabethan collar until his scabs could heal. We named him Hideous Henry—Hen, for short—and he lived with us for two years, until he developed a large tumor behind his left ear and we had him euthanized.

Killing is by far the most common form of human interaction with animals. The ways by which we kill animals and the meanings attached to these killings are as varied and diverse as the species of animals inhabiting this planet. Despite all this killing, a vast majority of people seem to believe that animals can suffer in ways that are morally important. Jonathan Safran Foer, for example, asserts that 96 percent of Americans believe animals deserve legal protection. (What we are protecting them against, I presume, is wanton cruelty.) That leaves a mere 12.8 million people who couldn't care less.

Killing may be our main form of interaction with animals, and all forms of killing are morally important, including how we kill "object animals" in slaughterhouses, research labs, and fur farms. But when it comes to our pets, the question of killing, and the broader question of how they die, takes on special importance. How should we feel about forms of intentional killing that are aimed to be in the best interests of our animal? Might it ever be morally appropriate to intentionally kill Ody? Can killing ever be an act of love? Assuming we can control at least some of the factors surrounding a pet animal's death, what might we do to make this death as good as possible?

THE BOOK, IN OUTLINE

Perhaps death matters to animals, and for reasons that might surprise us. Scientists studying wild animals have made the first few tentative steps toward what we might call animal thanatology. And those who study, care for, and live as companions with dogs and other companion animals are increasingly aware of their rich inner lives, and many believe that this richness may include some consciousness about death. What goes on inside the mind of a dog as he witnesses the death of a companion dog or as he himself draws his last breaths? Do animals grieve? Are animals aware of their own mortality? And are these questions answerable with anything firmer than conjecture? Because little research has been conducted on whether animals understand death, it is very hard to answer any of these questions with confidence. But although animal thanatology can offer us, at present, far more questions than answers, evidence and anecdote suggest that we have surprises in store. Animals likely have unique and fascinating ways of understanding dying and death.

Aging is intimately tied to death. But in addition to paying little attention to animal death, we suffer from a certain ageism in our attitudes toward animals. Biologists and ethologists categorize and study animals based on their age, recognizing that each life stage is physiologically and behaviorally unique. They study neonates, infants, juveniles, and adults. But there is no category for the aged, even though many animals, even in the wild, do live to be elderly and go through distinct physical and behavioral changes as they move beyond adulthood. Within the population of pets, the elderly is the fastest growing category. Despite increasing attention to their needs, prejudice against the old runs deep. Many elderly animals are euthanized simply because they are old or because their human owners don't have the patience or resources to adapt to their changing needs. And many senior animals languish in shelters. It seems to me that when we commit to owning an animal, we must commit all the way to the bitter end, as in a marriage. We can do many things to help them age successfully and to adapt to physical or behavioral changes. But as I know from living with Ody, doing our best is much easier said than done.

9

Given the popularity of pets in the United States, many people will share a portion of their life with an animal. Many will watch their animal age and eventually die, often after suffering from a painful decline. In deciding whether to hasten the death of an animal, the central consideration is usually pain. Although it may seem that pain is obvious, there are still many things that scientists don't understand about animal pain (or, for that matter, human pain). This is partly because pain is complex in its origins and manifestations, partly because pain is highly subjective and animals cannot communicate to us about their pain using our common currency of words, and partly because until recently people generally assumed either that animals didn't feel pain or that their pain didn't matter. Fortunately, the landscape of animal pain is undergoing dramatic change. Recognition and treatment of pain have greatly improved, and palliative care for animals is becoming more widely available and more effective.

The animal hospice movement, like a slow-moving glacier, is gradually carving out changes in how we care for our animals at the end of life. Hospice is neither, as some people mistakenly believe, a place to go and die, nor is it even so much a mode of treatment. Rather, it is a philosophy of care. It focuses attention on palliative treatments, maximizing quality of life, and shifting therapeutic priorities so that the emphasis is on care and comfort, not cure. Yet all is not easy and peaceful within animal hospice: lurking beneath the surface are deep moral disagreements about what constitutes a good death for an animal. Some believe that animals must be allowed to die a natural death, while others believe that the humane end point of animal hospice care is almost always euthanasia and that "natural" death is often far uglier than death under the needle.

Animal euthanasia is mired in contradiction. It serves on the one hand to release beloved pets from suffering and, on the other, to destroy millions of healthy animals whose only crime is being unwanted. It is certainly true that in human medicine there are plenty of opportunities to point out inconsistencies and absurdities. We clearly don't follow a simple maxim such as "all life has equal and unique and inestimable value" (else why would so many poor people die for lack of simple and inexpensive medical care? and why would so many pregnant women receive no prenatal care?). But with animals it is different. Not everyone values animals, and in fact many wear their contempt for animals quite visibly on their sleeve. And there are groups and individuals aplenty that either glorify (*Killing for Jesus*, the blog of a "Christian" hunter) or make light of (PETA—people eating tasty animals) or capitalize on the suffering of animals (such as the roadkill toys made to resemble various animals killed by automobiles each year, including Twitch the raccoon and Grind the rabbit). Yet perhaps more troubling than outright contempt for animals is the nonchalance that feeds the tremendous problem of homeless animals and the plight of the millions of animals killed each year in shelters and pounds across the nation, to say nothing of the billions of living creatures killed in slaughterhouses, research and product-testing labs, and fur farms. How do we embrace euthanasia as an act of mercy for suffering animals, while challenging its use as a tool for easy disposal of unwanted or fractious animals?

What remains after an animal has died? Well, a body, for one. Some people choose to bury their pets in the backyard, but it can be tricky, depending on the neighbors, and is, in many places, illegal. Another option is a pet cemetery or the increasingly popular choice: cremation. Of course many other more ephemeral things remain as well—our grief, our memories, perhaps some guilt over things done or not done. The time we lost our temper with our dog and smacked him over the head with a newspaper, or the fact that we put off making a vet appointment because other things took priority. Aftercare for companion animals is gaining attention, expanding the range of options for people wanting to memorialize their animal and providing options for those who believe that a deceased animal's body should be treated with care and respect.

I want to celebrate animal life, and this includes celebrating the fact that their death is meaningful, both to us and to them. Many of us love animals deeply and choose to welcome them into our homes, and as their companions we try to take full moral responsibility for our relationship with them. One of my early ethics professors, Ed Freeman, always started his lectures by saying that ethics is for good people. It isn't for the crooks and liars; it is for those of us who are trying to do right in the world. I couldn't agree more. This book is about the kinds of quandaries that good people find themselves in with regard to their aging or dying animals.

It is also about embracing a certain level of moral failure. I have learned from my experiences with Ody that doing the right thing is not so easy. I could have done more for him. There are things I did wrong, things that I failed to do simply because there were too many other demands on my attention, and things I neglected to do because I didn't know enough. I have sometimes taken the path of least resistance and have even felt that my life would be easier if Ody would just die already and be done with it. When it comes to our animals, we need large measures of compassion and a little bit of self-forgiveness.

THE SIXTH FREEDOM

Read around in the animal rights and animal welfare literature and you will soon come across one mention or another of the "five freedoms." The five freedoms originated in Britain during the 1960s, when the welfare of farm animals came under careful discussion. The *Report of the Technical Committee to Enquire into the Welfare of Animals Kept under Intensive Livestock Husbandry Systems*, otherwise known as the Brambell Report, established minimal welfare standards for agricultural animals that included the freedom to stand up, lie down, turn around, groom, and stretch one's limbs. This may not sound like much, but at the time this affirmation of basic animal needs verged on the revolutionary. It was an outright admission that we have a moral responsibility to consider *their* needs.

Recognizing that the ability to turn around in a stall or cage was inadequate to true animal welfare, the UK Farm Welfare Council revised the five freedoms in 1993 into their current form.

1. Freedom from hunger and thirst
2. Freedom from pain, injury, and disease
3. Freedom from discomfort
4. Freedom to express normal behavior
5. Freedom from fear and distress

Of course, and as the Farm Welfare Council was careful to point out, these are not meant to represent attainable goals but, rather, ideals. A pig raised for bacon will never be free of fear, distress, pain, and discomfort and will never fully be able to express normal "pigness." Still, we can work toward the greatest possible freedom for animals, within the constraints of our own needs.

Although these standards of animal welfare were established for agricultural species such as cows and pigs, the five freedoms have become a common baseline for discussing the welfare of all animals held in human confinement, from zoo animals to dogs in shelters to mice in laboratory cages. Even those of us who believe we are true animal lovers, attuned to the needs of the animals with whom we share our homes, will do well to pay attention to these five freedoms. We might surprise ourselves by discovering that we could do much better.

I also suggest that we add a sixth freedom: freedom to die a good death. A good death is one that is free of unnecessary pain, suffering, and fear; it is peaceful; and it takes place in the presence of compassionate witnesses. It is, above all, a death that is allowed its full meaning. Is it strange to say that we might desire for the death of an animal to be meaningful? Can the ultimate act of obliteration have meaning? Yes. It can and does. Death is obviously meaningful to the one who dies. It is like the final cadence at the end of a piece of music, bringing the music to its necessary harmonic resolution. But perhaps it is especially so to those who survive. Death affirms the value of life, and if we are to value animals, we must value their deaths.

TO DIE LIKE AN ANIMAL

"To die like an animal" is a phrase of common understanding. Consider this story, recently recounted in a newspaper: A junkie named Michael Faulkner was sentenced to three and a half years in prison for the crime of leaving a man to die "like an animal." Mr. Faulkner had injected a fellow junkie with heroin, but the man had an adverse reaction to the drug—probably because he already had high levels of alcohol in his bloodstream. Instead of calling the police or an ambulance, Mr. Faulkner left the man, dying and alone, behind a local bar. To die like an animal, in the idiomatic sense, is to die in a way that is beneath dignity, to be forgotten or left, deliberately, to suffer, to die in a way that none of us desire. It is to die a bad death.

It would be nice to live in a world where "dying like an animal" signified a peaceful, respectful, and meaningful death.

The Ody Journal

I began keeping the "Ody Journal" in the fall of 2009, and wrote almost every day until Ody died a little more than a year later, just after Thanksgiving. I intended to keep writing after he died, as a way to process the grief. But I found that I couldn't. So it all stops abruptly, as his life did.

At the beginning, the entries tend to focus on Ody's antics—mostly the mischief that he seemed to seek out, his quirks, his daily business. As the story continues, Ody's aging and health problems increasingly take center stage. It was uncanny, the way this book unfolded alongside Ody's dying. When I began writing, Ody was a grumpy old man, and death hung far out on the horizon, like a sun heavy with the day's work. By the time I reached chapter 5, on hospice, Ody was in serious trouble. I finished chapter 6, "Blue Needle," on the day before Ody's final appointment with the vet, and I struggled through the final chapter as I grieved.

The actual "Ody Journal" filled up two notebooks. I've left out some entries here. Particularly toward the end of Ody's life, it all became tedious, with lots of barking in the middle of the night, accidents on the floor, and discussions with vets. Other than cutting, I haven't changed the journal—it is raw and rough, the way the house looks when you aren't expecting company.

THE CAST

Ody: You'll meet him soon.
Maya: Our German shorthair /English pointer mix. She is seven as the story begins. Maya has been a part of our family since she was about twelve weeks old. Her mother, Ocho, lives several miles from us, out in Hygiene,

Colorado; her father, Buzz Lightyear, belongs to some friends who also live close by. Buzz was bred for bird hunting, and Maya has incredible hunting instincts. She goes on point multiple times a day, for squirrels in the backyard and birds in the driveway and sometimes a paper bag or a clump of leaves. Her nickname is Bird Brain. She is not all that smart, but she is one of the sweetest dogs I've ever known. Maya loves Ody and protects him like a mother hen.

Topaz: Our red heeler/border collie mix. We adopted him as a ten-week-old pup, with the intention that he would be Sage's dog to raise and train. Unfortunately, he bonded to me and, as happens with heelers, became a one-person dog, attached to me more firmly than my own shadow. He is about a year old as the story begins. Topaz is on a mission to rule the house and has gradually undermined Ody and Maya alike, making incremental power grabs whenever the opportunity arises. You can almost see him scheming. Sadly, he is the bane of Ody's existence. His nickname, when he is being nice: Wazzy. When he is being mean: Cerberus, because he jealously guards anything of potential value (me, food, toys, bones, the office, the bedroom, the kitchen, the doorway to the house) and basically makes life hell for Ody.

Chris: My husband and partner in this journey.

Sage: Our daughter. Eleven years old as the story begins.

SEPTEMBER 29, 2009–JANUARY 15, 2010

SEPTEMBER 29, 2009

Ody opened the pantry and took out the box of Emergen-C vitamin drink packets. He tore up the box, but didn't seem interested in the powder.

OCTOBER 5, 2009

He sat out in the backyard and barked at the door for about an hour, until the neighbor got fed up and called me on my cell phone. I was in Fort Collins at the cross-country meet with Sage's team. Incredibly, the dog door was open, and there was no evident reason—no physical reason—that Ody couldn't just go inside.

He opened the pantry and pulled out the bag of potatoes. Apparently they didn't appeal.

OCTOBER 6, 2009

Two bags of rat food, half-eaten, are strewn all over the living room floor.

OCT. 10, 2009

I splurged yesterday and bought a pretty white ceramic flower pot at the Flower Bin. I set it on the dining room table, so that I could put a plant in it the next day. Unfortunately, Sage laid some treats on the table last night, for training the dogs, and then forgot them. Ody followed his nose, and the treats and the pot must have been too close together. We woke early this morning to a loud crash and found the pot in pieces on the floor. All the treats had been carefully picked out of the rubble.

OCTOBER 12, 2009

Ody peed on Sage's floor, right in front of me. Just squatted down, looked over at me, and let it rip.

Today Ody experimented with flour. He didn't seem to enjoy the whole wheat, but the white must have been tasty. A trail of flour led from kitchen pantry into living room, then an explosion of white all over the wood and carpet.

OCTOBER 20, 2009

Ate umpteenth canister of fish food. He chewed through the plastic container, and spread Tetramin flakes all over Sage's room. Everything smells fishy now—my hands, my clothes, the air, the vacuum—and little fishy particles are ground into her carpet. Yuck.

This has been life with Ody: every single day, if I dare leave the house, I come home to some kind of mischief and some mess to clean up.

Some dogs have no problem being left at home, or so my friends with dogs tell me. These lucky dogs remain relaxed and simply sleep or work sentry duty. Leaving Ody, even for quick runs to the grocery story, has always been an ordeal. He spends the time I'm gone in a kind of schizophrenic mania, fluctuating back and forth between a state of near panic and rabid food mongering. Both impulses are destructive, and every time I return home to Ody it is with trepidation. I crack open the door slowly, and hold my breath. As Ody dances and jumps around me in wild delight, I survey the damage, add up costs, and calculate how long the cleanup will take.

Long ago, we developed the Ody Algorithm, which would need to be applied every time we had to leave the house. Step 1: can Ody possibly come? If not, proceed to step 2: all edible items off the counter? Pantry door firmly closed? Rat food and fish food stored on a high shelf? Backyard electric fence turned on? Backdoor screen latched? Basement door closed to protect furniture? Bedroom doors closed to protect bedspreads? Couches piled with chairs to protect cushions? And so on.

When Ody was a puppy, we tried the crate approach, which is supposed to help dogs feel safe and remain calm when their owners are away. We kennel trained Ody just as the monks of New Skete instruct the readers of their training guide, carefully adjusting upward the amount of time Ody was in his kennel. But Ody never behaved the way the lovely shepherds of New Skete do. He hated the kennel from the start, and his hatred only grew more intense over time. He would pant and salivate the entire time he was kenneled, and when I let him out he would be soaked to the bone in his own slobber. He would dig frantically and bite at the sides and openings of the kennel, leaving the metal deeply grooved and his mouth bloody. Once I came home from work to find Ody completely covered, nose to tip of tail, in his own diarrhea, having become so anxious in his confinement that he lost control of his bowels. After that, we permanently retired the crate.

Ody is what you would call, with ironic understatement, high maintenance.

He suffers from many variants of anxiety, including (at the least) separation anxiety, thunderstorm mania, and generalized anxiety. To deal with his psychoses, we've taken Ody to countless vets, including specialists in veterinary psychiatry. They offered various prescriptions that would without fail cure Ody: desensitization, retraining, drugs.

Desensitization didn't work. For example, we tried to help Ody with his fear of thunderstorms and loud noises. We searched high and low and finally found a CD recording of thunderstorm sounds and would play it quietly in the background while feeding Ody treats and playing with him. Gradually we increased the volume. Ody wasn't the least bit worried by the fake storms and happily gobbled up his hotdogs and Milk Bones. Ten minutes after an excellent desensitization session, however, Ody could walk outside, spot a cloud in the sky, and immediately crumble into a panic. When in the grips of an anxiety attack, he would begin to pant and pace and tremble. I remember many a stormy night in Nebraska: Ody would try to get as close to us as he could, which meant panting and quivering and pacing back and forth on the bed and, specifically, on pillows—and our heads—while the storm raged. Ody is so afraid of thunderstorms that he refuses to go for a walk or even go out in the yard if the sky is overcast.

Psychotropic drugs never helped either. We tried Prozac, Xanax, Ativan, and a collection of other medications that are supposed to help canine anxiety. These either left Ody so drugged that he was no longer himself, elevated his liver enzymes to dangerous levels, or had no effect whatsoever.

Retraining never even crossed our minds. One doesn't train an Ody out of his anxiety.

Only in his extreme old age has leaving Ody become somewhat less worrisome. He is mostly blind and almost completely deaf and is usually asleep, so he is largely unaware of our comings and goings. His physical capacity to inflict damage on his surroundings is radically diminished (not including, of course, damage of an excremental nature). He has no teeth to speak of, and his hind end is too weak to allow him to jump the fence or crawl high onto the counters. He just cannot dig, tear, chew, climb, or scratch as he used to. Still, he manages what mischief he can. He still gets into food on the counter if it is near enough for him to reach and on the low shelves in the pantry.

I thought I should record some of Ody's nicknames. Do all dogs earn so many monikers?

- Destructo-Dog: No explanation needed.
- Odiferous: He loves to roll in dead stuff and anything related to feces.
- Odious: Sometimes he is.
- Boney-Man: A special old-man nickname. His elbows and knees and hips have gotten all pointy and sharp.
- Toothless Wonder: Another old-man name, because he can eat unbelievable things with no teeth.
- Psycho: Not very nice to make fun of his mental problems, but sometimes we can't help it.
- Red Man: For his color, of course. But with double meaning: Ody grew too curious at a friend's stable, and after wandering into the pasture he was chased, tail between his legs, by a mule named Red Man, and he had to hide under a truck bed. I couldn't get him to come out for a long, long time.
- Blast-ended Skrewt: For his inscrutable nature, combined with an odd behavioral trait: his back end puffs out in a most disturbing way when he needs to go to the bathroom. Credits to J. K. Rowling.

· Buffalo or Buffy: Short for water buffalo, his totem animal. Ody doesn't swim (never figured out how), but he loves to wallow in chest-deep water.

NOVEMBER 11, 2009

We took Ody with us on a trail run. Usually these days we leave him at home because he can't go very far and seems stressed out trying to keep up, even on a slow hike. Sage was with us, and we figured she wouldn't want to run far, so we trundled Ody and the other dogs into the back of the car. Ody looked so happy, frolicking in the snow. It was quite cold, which suits him. I swear that Ody was grinning the whole way—I know it just looked like a grin because he was panting, which pulls his catfishes back. ("Catfishes" is our technical name for Ody's lips; that jaggedy fleshy bit next to his teeth.) But I want to think that he was really smiling.

It was bittersweet for me, the joy of seeing Ody out running in the forest, which he has always loved, and the heartbreak of watching him struggle. It took all his attention and focus and energy just to keep his feet under him and stay within twenty or thirty feet of the rest of us. He began to pant almost immediately and kept up with the heavy breathing for the whole run.

He slipped over and over in the snow. On the descents, he was almost bobsledding down the trail on his butt because his legs kept collapsing. Once, he fell through a hole in the snow and took a nose dive into a snowbank.

Ody seems to very gradually take stock of his changing physical capacities. About a year ago, when his rear legs were starting to become weak and stiff, he would still try jumping into the back of the car, and he had unrealistic expectations about succeeding. I'd say his odds were only about fifty-fifty, but he had the courage (or stupidity) to try, over and over. Sometimes his front paws made it in, but his hind legs didn't quite follow. After a mighty struggle, he might be able to worm his way up. Or, just as often, he would fall backward out of the car and wind up flat on the cement. I had to make sure I was poised right behind him before he decided to jump. On a few occasions, he decided to jump just at the moment I leaned down to pick him up, giving me, in this order, a fat lip, a bruised cheekbone, and a bloody nose. Letting him out of the car was even scarier because Ody would take a flying leap and oftentimes his legs wouldn't catch his weight and he'd land on his nose and roll over onto his back, where he would be trapped, like an up-ended beetle. Now he's wised up, and he simply refuses to go near the car. I have to carry all seventy pounds of him from the door to the car and hoist him into the back. And I have to pry him out once we arrive where we're going. I also have to open the hatch with utmost care so he doesn't accidentally tumble out or get knocked out by one of the other dogs, as has happened a few times.

NOVEMBER 22, 2009

Ody is teaching me a great deal about patience and compassion. I try really hard not to feel annoyed by his sit-outside-the-door-and-bark stunt. I guess this is what it is like to care for an older person—we try to put their behavior in perspective.

NOVEMBER 25.

The barking is really a problem. After thirteen years of happily using his dog door, Ody has suddenly stopped going in on his own. He can go out the dog door, but he will not go in. And so, he goes out and must then bark in order to be let back inside. All day, over and over, he goes out. And then barks. It feels as if I'm getting up from my computer every five minutes to let him in.

DECEMBER 6, 2009

As Ody was eating dinner, his back left leg kept sliding inward, as if his muscles weren't cooperating. He'd correct, and then it would happen again.

He almost fell while climbing the stairs from the basement. His back legs began to collapse. Luckily I was close by and heard him slipping. I ran down the stairs and hoisted him up. He was breathing heavily and had a terrified look in his eyes.

He has seemed more sociable the past few days, actually seeking us out and lying in the same room. But he is also very restless. He wanders around the house, and then wanders into the room where I'm working or reading and stands there, looking perplexed. He goes outside, and then barks to come in. Goes back out, barks. Wanders briefly into the room, then back out.

Do dogs understand aging? When we force Ody to go on a hike or jog in the woods, does he feel sad because he can't run around with the other dogs? I think probably not, but I'm not sure.

We took him to our secret Left Hand Canyon trail—it was snowy, and he had a hard time, slipping off balance, and getting obviously tired by the physical exertion. His back legs were sliding out from beneath him. Every ten or twenty steps, one of his back feet would roll over and he would land on top of his toes (the vet calls this "knuckling"). But did he still have fun? How could I know?

Most of the dog books I read seem to be about happy dogs. The owners may not be so happy—until they come to know one particular canine who turns their world around. But the dogs themselves are joyful and sweet and teach us lessons about embracing with unrelenting optimism everything that life throws at us. Even Marley—troublemaker extraordinaire—was an upbeat kind of guy. But what about the cup-half-empty kind of dog? The worrier? The dog in existential crisis?

DECEMBER 20.

More barking. Morning, noon, and night. Driving me crazy.

DECEMBER 31, 2009. STEAMBOAT SPRINGS.

I believe Ody has eaten at least one bedspread in a Steamboat hotel, perhaps several. So the place is dear to him.

Since this is our last trip of the year, I worry that it will be Ody's last Steamboat trip ever. Vacations are becoming so hard. I can't bear to leave Ody at Pansy's,♦ but our usual pace of activity is hard on the old guy. He has trouble walking in deep snow, and there is so much snow in Steamboat that it will be hard to avoid.

We took him up the Spring Creek trail, which gets a lot of traffic and tends to be packed down in between snowfalls. And he did okay. But I had to leave him in the car while I took the other two dogs snowshoeing at Fish Creek and this morning when we skied at Rabbit Ears pass. It felt so unfair to Ody.

Ody can't go up and down the stairs to our hotel—they are made of that grated metal you find in ski resorts, meant to keep people in boots from slipping. But the grating is hell on dog paw pads. Even Topaz and Maya hate walking on the stairs, but they can leap down through the snow banks

♦ Pansy's Canine Corral is where the dogs go when we have to leave town and they can't come. It wouldn't be right to call Pansy's a kennel; it is a home-away-from-home for dogs. All the dogs hang out together, in a huge fenced-in area on Pansy's farm. They sleep either in the carriage house or in the enclosed porch or—especially for Ody—in the basement. Ody has an old armchair that he loves.

to the side. We have to carry Ody up and down, each time we leave. And he seems uncomfortable, in general. He has a hard time riding in the car now (panting, standing the whole way), and even in the warm hotel room he seems uncomfortable. Much of the time he stands or lies upright by the door (sitting is hard for him; he has no sit any more). Maybe he is afraid to really come in because of Topaz. And it seems so sad at night, when Maya and Topaz are cozily curled up in the beds with us and Ody lies alone on the floor. One of Ody's favorite things used to be sleeping in the bed with us. He would burrow down under all the covers, and he was typically the last one up in the morning.

I lifted him onto the bed this afternoon and now he's sprawled out, snoring loudly and engaged in some very active dreaming, legs not just twitching but convulsing.

He seems so remote now—he doesn't hold my gaze the way he used to, and he seems fairly uninterested in affection.

MID-JANUARY, 2010. HOME.

Whenever our friends Liz and Craig come to visit they mourn over Ody and how much he has changed. They have known Ody since he was a crazy two-year-old. Liz says, "He's just not there anymore," "He's gone," "He's just . . . vacant." And it's true. He is not as he was. I don't think that he is gone; he has simply receded. His "person" dwells somewhere deep down.

I get to remembering Ody in his fullness: running wild in the tall grasses at Chalco Park in Omaha, sometimes scaring up a pheasant, which would send him into an ecstasy of excitement; racing through Frick Park in Pittsburgh, down among all the other people and dogs who would gather on a weekend morning; making his way through a playground full of children or through the middle of a soccer game—saying hello to each and every person.

I remember a particular day with Sage and Ody, walking at the park off of Beechwood Avenue in Pittsburgh where we spent so many hours and where Ody would race around playing with the other dogs. On this day, Ody ran up to a little boy and his father. The boy was holding a hotdog in one hand. Ody ever so gently and ever so quickly put his mouth around the hotdog and gulped it down, before I could blink or cry "ODY! STOP!" The boy was too stunned to cry. I apologized profusely, offered to buy the boy a new hotdog, and when this offer was refused, turned around and dragged Ody and Sage out of the park as quickly as I could.

I wonder if Ody's barking means that he is lonely. I try to recast barking as "vocalization" because it sounds more clinical, as if maybe there is a behavioral explanation for it. The last thing I want to do is get annoyed. So I do some cognitive behavioral therapy on myself. I "read" his vocalizations as one of the few means of communication open to him, as a way for him to reach out to us. And then his barking just makes me sad. Ody is the most social dog I've ever known. Now he lives in isolation. I see him trapped in his bed under the piano, while Topaz guards me, the kitchen, and the door. Ody always—for thirteen years—slept in the bed, under the covers. Now he sleeps alone.

I hear Ody barking in the night, even when he's not.

2

Into the Open

With all its eyes the creature sees
the Open. Only our eyes are
as though reversed, and placed all around them
as traps, encircling their free exit.
What *is* outside we know only from the animal's
countenance; for already we turn around
the young child and force it to see backwards,
see form, not the Open that's
so deep in the animal face. Free of death.
It, only we see. The free animal
has its demise always behind it
and before it God, and when it moves, it moves
in eternity, as the fountains move.

RAINER MARIA RILKE, excerpt from "The Eighth Elegy"

Like so many animals conceived by poetic imagination, Rilke's animals are free of death. They die, of course. But they live without dread of the blackness that threatens to envelop every living creature. They peer only into the Open, into that limitless blue sky of possibility that life spreads before us. "Nor dread nor hope attend / A dying animal," Yeats writes. "A man awaits his end / Dreading and hoping all; / . . . He knows death to the bone - / Man has created death." Rilke and Yeats may be right: animals may not dread death as we do. And perhaps humans, in an important sense, have created death. But this is certainly not all there is to say about animals and death. It is tempting to say, simply, that animals cannot understand death because they are not human beings. But this is a silly tautology. Of course animals don't understand death as human beings do. But does this mean that animals don't understand death, even in their own animal fashion? We need to ask, "What does death mean *for an animal*?" Or, better yet, "What might death mean to this particular animal, given its unique cognitive and emotional capacities, its social attachments, its life experience, and its idiosyncrasies?"

Animal death matters from a scientific perspective because research into death awareness relates to broader questions about animal cognition, emotion, and social behavior. And, it matters from a moral point of view. We can be very cavalier about how animals die—and about how we kill them. Yet given that aging and dying are among the most significant life events for all creatures, we might give some thought to their ends, particularly for those animals in our care, whose lives we control and whose deaths we orchestrate. If a good death is what we're after for our animals, we need to think about what it is that animals are actually experiencing as they die or as those around them die. This understanding will also help us avoid inflicting bad deaths on the animals under our care.

We seem to be on the cusp of a significant renaissance in our thinking about animals. They are not dumb brutes, not by a long shot. We have learned over the past two decades that our easy assumptions about animals—for example, that they do not feel pain, cannot use tools, possess no capacity for empathy—are continually being upturned by science. We have also learned that our scientific language often falls a little flat: we may feel a cool scientific detachment using terms such as "nociception" and "affective valence," but we may also find that this language feels a bit inauthentic. It is of course true that we can ascribe too much: "little Mitzy felt so guilty for chewing my brand new slippers!" and, my favorite, "Ody is so lonely when I leave him home." But we can also ascribe animals too little. We need to live on the edge a little and allow ourselves to be "unscientific" now and then. We need to allow the behavior of animals, particularly those we know intimately because we share our lives with them, to come to life through language that we can understand emotionally and sympathize with: pain, contentment, love, grief, terror, sorrow, longing, and joy.

ARE ANIMALS AWARE OF DEATH?

This question is very hard to answer because little sustained scientific research has been conducted on death awareness in animals, or what might be called—for those who appreciate academic-sounding labels—animal thanatology. What we do have are some tantalizing clues that provide reason to at least take the question very seriously. Clues come in from various parts of the animal world—some from undomesticated animals in the wild; some from wild animals in captivity; and some from domestic animals, especially our companion animals, and from dogs in particular. We know enough, I'd say, to posit that at least some, and perhaps many, animals have death awareness of some kind. "Animals" is a heterogeneous grouping, of course, and what we might observe in one species cannot be assumed to be present in others.

I hope there will be sustained research into this area, including careful work to distinguish those questions that are open to scientific investigation from those that are not. For those that are not, it doesn't mean we can't ponder them. Philosophers have been working productively for several millennia on the question of what a good life is. Exploring such intangibles just means that we must use a different set of investigatory tools and seek different kinds of answers.

DEATH-RELATED BEHAVIOR IN CHIMPANZEES

In the spring of 2010, a study by James R. Anderson, Alasdair Gillies, and Louise C. Lock of the University of Stirling reported on observations of a small group of captive chimpanzees in Scotland. The researchers took video recordings of three chimpanzees reacting to the dying of a fourth member of their group, an elderly female named Pansy. The chimpanzees groomed Pansy just before her death. Just moments after Pansy finally died, Chippie (the male) jumped onto the platform in an aggressive display, leaped into the air, and brought both hands down and pounded her torso. After her death, the other chimpanzees closely inspected Pansy's mouth and manipulated her limbs, perhaps testing for signs of life. They removed bits of straw from her body. Pansy's daughter Rosie stayed with Pansy's body almost continuously on the night after she died. Following Pansy's death, all three chimpanzees slept fitfully and moved around a lot in the night. For several days following Pansy's death, the others avoided the platform where death had occurred. And for several weeks they were subdued, lethargic, and ate less than normal. Beyond simply describing the reactions of the groups to Pansy's death, the researchers give their own interpretive spin, suggesting that the group's responses parallel, in striking ways, human responses to the death of a close relative: predeath care, inspection of the body for signs of life, an after-death vigil, cleaning the body, and avoiding the place where death had occurred. This interpretation strikes me as a tad overreaching, since we cannot know or even infer what the chimpanzees were feeling or what motivated the behaviors. And the behavior of one group of captive chimpanzees cannot be generalized to wild chimpanzees or even other captive populations. But even without the spin, the study is worth attention because (a) the animals clearly reacted to Pansy's death, and this is really interesting; and (b) by taking the question of death awareness in animals quite seriously, this study is a first step in establishing animal thanatology, and in this particular case "pan thanatology," as a viable subject of research. (Kudos to the journal *Current Biology*, which published this study, for going out on limb.)

THE THINGS THEY CARRIED

Another death-related chimpanzee study involved observations of a group living in the forests surrounding Bossou, Guinea. During 2003, a spate of respiratory illnesses took the lives of several chimpanzees, including two infants. Researchers noticed that chimpanzee mothers carried the remains of their dead infants around with them, sometimes for so long that the remains became dry and mummified. During the time they were carried, the bodies of the dead infants first swelled and then dried out and lost their hair. In photos of the animals, the infant corpse is flat and stringy and drapes across the mother's back, an empty, leathery backpack. Limbs hang down, like spidery straps. The mothers carried the bodies everywhere, often by gripping a leg or hand. They groomed the bodies and chased away flies. Other related and unrelated chimpanzees in the group "attempted to touch, poke, or handle" the bodies.

In other chimpanzee communities, the researchers note, corpses of dead infants are often snatched away from the mother by others in the group. The bodies are often treated violently and are sometimes even eaten. But in the Bossou community, there appears to be a particular attitude toward infant corpses, and the researchers suggest that there may be some observational learning occurring among this group of animals.

Did the chimpanzee mothers understand that their infants were dead? The researchers leave this question unanswered. The only firm conclusion they draw about corpse-carrying behavior is that it testifies to the depth of the mother-infant bond among primate species. In a *New York Times* story about death awareness in animals, science writer Natalie Angier takes a skeptical view of the corpse-carrying research. She concludes that although chimpanzees seem to be aware of death, they generally act as if it were no concern of theirs, and in support she quotes anthropologist Michael Wilson, who has studied chimpanzees at Gombe: Chimpanzees, he says, are different from us in terms of what they understand about death and about the difference between the living and the dead. Although infant chimpanzees will obviously mourn when their mothers die, adult chimpanzees, he says, take a laissez-faire attitude toward the dead.

It might be worth pointing out the obvious: this is also true for the human animal. We do not bother over every single death, and some deaths matter much more to us than others. Some deaths, in fact, affect us not at all. I may be aware of another's death and think about it but not outwardly alter my behavior in any way. So looking at me from the outside, you might label me laissez-faire, too.

MORE DEATH-RELATED BEHAVIOR IN WILD ANIMALS

A 2009 *Daily Mail* headline read, "Is This Haunting Picture Proof That Chimps Really DO Grieve?" The photo was shot in the Sanaga-Yong Chimpanzee Rescue Center in Cameroon, West Africa, and shows a group of sixteen chimpanzees crouched just behind the wire fence of their enclosure. They watch intently as one of their community members—a forty-year-old named Dorothy who had died of heart failure—is wheeled past.

Just a few months earlier, another story about animal grief had captured media attention. Gana, an eleven-year-old gorilla at the Munster Zoo in Germany, had clutched her three-month-old baby after its death, refusing to allow zookeepers near the body. The pictures of Gana are heartbreaking: she holds her baby Claudio's limp body high up above her, as if toward the heavens. It is very hard to look at these photos and not feel your heartstrings being vigorously tugged, and this makes it all the easier to think of these as images of grief (particularly when the captions read "chimpanzee funeral" and "grieving mother"). So we must view these with caution—we do not (and cannot) know that these animals are feeling grief. Nevertheless, aren't you curious to know more?

In *Through a Window*, Jane Goodall describes the behavior of a young chimpanzee named Flint on the death of his mother, Flo. Because Flo was so old when she had Flint, she didn't have the energy to wean him and he remained dependent on his mother. Goodall says, "Never shall I forget watching as, three days after Flo's death, Flint climbed slowly into a tall tree near the stream. He walked along one of the branches, then stopped and stood motionless, staring down at an empty nest. After about two minutes he turned away and, with the movements of an old man, climbed down, walked a few steps, then lay, wide eyes staring ahead." Flint sank into a depression. He became lethargic, stopped eating, and fell sick. "The last time I saw him he was hollow-eyed, gaunt and utterly depressed, huddled in the vegetation close to where Flo had died . . . The last short journey he made, pausing to rest every few feet, was to the very place where Flo's body had lain. There he stayed for several hours, sometimes staring and staring into the water. He struggled on a little further, then curled up—and never moved again." Did Flint die of grief?

Other scientists who study animal behavior have also observed grieving. Konrad Lorenz, for instance, described grieving in a greylag goose. "A graylag goose that has lost its partner shows all the symptoms that John Bowlby has described in young human children . . . the eyes sink deep into their sockets, and the individual has an overall drooping experience, literally letting the head hang. . . ."

Death-related behavior in elephants has been widely reported. Zoologist Iian Douglas-Hamilton believes that elephants have a general awareness of and curiosity about death. They will gather around the body of a dead herd member, gently touching the body with their trunks and feet, often standing vigil for days. Elephant researcher Cynthia Moss writes,

Even bare, bleached old elephant bones will stop a group if they haven't seen them before. It is so predictable that filmmakers have been able to get shots of elephants inspecting skeletons by bringing the bones from one place and putting them in a new spot near an elephant pathway or a water hole. Inevitably the living elephants will feel and move the bones around, sometimes picking them up and carrying them away for quite some distance before dropping them. It is a haunting and touching sight and I have no idea why they do it.

A study of tool use in African elephants found that they will sometimes put food in the mouth of the dead, pack the wounds of the dead with mud, and will bury their dead under vegetation. And biologist Joyce Poole writes of elephants, "I have observed a mother, her facial expression one I could recognize as grief, stand beside her stillborn baby for three days, and I have been moved deeply by the eerie silence of an elephant family as, for an hour, they fondled the bones of their matriarch."

According to a report by the Cornell Lab of Ornithology, yellow-billed magpies react to a death by descending on the carcass and hopping around and squawking. Ethologist Marc Bekoff observed the following behavior among a group black-billed magpies: "One approached the corpse, gently pecked at it, just as an elephant would nose the carcass of another elephant, and stepped back. Another magpie did the same thing." Bekoff continues: "Next, one of the magpies flew off, brought back some grass and laid it by the corpse. Another magpie did the same. Then all four stood vigil for a few seconds and one by one flew off." Biologists don't know the function of these behaviors but some, including Bekoff, have described this as "funeral behavior."

DEATH AWARENESS IN COMPANION ANIMALS

We find common references to death awareness in companion animals. "There can be no doubt," proclaims veterinarian Michael Fox, "that animals possess some understanding of death." He is speaking here particularly about dogs. I cannot tell you how many stories I have read and heard about dogs who, in one interesting way or another, seemed to be aware of and react to the death of a companion—enough to fill up this entire book. And I have my own story, which will come in time. These tales reflect a wide range of human observations: sometimes dogs react to death by howling or whining, sometimes by seeming to become depressed and listless, sometimes by searching or standing vigil for the missing companion, sometimes by curling up next to the dead body. If you have dogs (or cats, for that matter), or know someone who does, you likely have a story of your own.

Fox offers extensive coverage of this topic in his *Dog Body, Dog Mind* in a chapter titled "How Animals Mourn and Express Grief." "Animals, in their grief, and in their longing for the return of the loved one, show us that the nature of love is an inborn quality that we share with them." But outward appearances, he says, can be deceiving. Animals may not outwardly express their grief in ways discernible to us. Sometimes the first response of an animal is acute grief and crying. Some animals show no initial reaction to the death of a companion (human or animal). Later, though, they may begin to search for their loved one, becoming more and more apprehensive and vigilant. Some dogs will show signs of depression, loss of appetite, listlessness. Some will vocalize; others will grow quiet. Some will become clingy; others withdraw.

The Companion Animal Mourning Project, carried out by the American Society for the Protection against Cruelty to Animals, found that two-thirds of all dogs in the study exhibited four or more noticeable behavioral changes after the death of a canine companion. More than a third of dogs ate less than usual after the death of a canine companion, 11 percent stopped eating altogether, and almost two-thirds vocalized more or less than normal. Many changed the location or pattern of their sleep. Some became more clingy, others more distant.

Fox offers a number of stories his clients have sent him of animals responding to the death of a companion animal. I have, in the course of researching this book, heard more personal anecdotes than I can possibly share. Too many, I think, to simply write off as coincidence or projection. And many of the stories come from veterinarians, who we would expect to have a certain threshold of scientific sensibility.

Two examples follow.

A woman named Mandy told me about the reaction of her Yorkie, Gizmo, to the death of her miniature pinscher, Diamond. Gizmo, who had lived with Diamond since they were both puppies, went with Mandy and her family to the vet to have Diamond euthanized. Gizmo didn't pay much attention while Diamond was put to sleep, but once Diamond was dead, Mandy brought Gizmo over to the body. Gizmo sniffed the body up and down and then sat down on his haunches. Mandy said, "I swear to you, he stared at Diamond and furrowed his brow."

Shawn related her experience with two Rottweiler sisters, Delilah and M. Like Gizmo and Diamond, these two had lived together since they were puppies, and they died within a couple of months of each other. M was the first to go. She was very sick, and a vet came to the house for the euthanasia procedure. Just after M died, Delilah jumped up on the bed where her sister was lying and licked her face carefully—not a typical behavior for her. Then she jumped down and ran off. She didn't seem depressed but did change her behavior: she began eating only from her sister's bowl and would sleep in her sister's place.

What I find most intriguing about these and other stories like them is both the variety of reported behaviors and the frequency with which people observe behavioral changes in one pet after the death of another. I wonder whether companion animals have an increased sensitivity to the death of a companion because of the environment within which they live and the extraordinarily close bonds that can form when only two or three animals live together. Or it could simply be that we have so many stories because so many of us have the opportunity to closely observe animals in our homes. We also perhaps feel more liberty to label behaviors in our pets as humanlike (sadness, depression, loneliness, grief), since the boundaries between people and pets are much fuzzier than between people and wild animals. Perhaps these stories are not so much about animals as they are about how humans see animals.

Let's assume for a moment that animals do, indeed, grieve. If it makes you more comfortable, you can insert some scare quotes around animal "grief." Despite important continuities, animal grief is not equivalent to human grief, and because of the subjectivity of emotion we will never understand or experience animal grief. Furthermore, grief does not prove an understanding of death. We don't really know *why* they grieve. It could very well be that grief is a response to the loss of companionship. All we know is that for them, as for us, grief is a form of suffering and that it has psychological and quite physical manifestations. We also know that grief is highly individual. Some grieve not at all, some grieve quite functionally and for a limited time, and some are devastated by grief.

We've talked so far about how animals react to the death of another animal. What about their own death?

WHAT DO ANIMALS UNDERSTAND OF THEIR OWN DEATH?

A swan song is a final dramatic appearance or accomplishment, one last and most magnificent flowing forth. The idiom comes from an old belief that the so-called mute swan (*Cygnus olor*) is completely silent until just moments before it dies, at which point it erupts into a final beautiful song. It turns out that these swans are actually neither mute—they make snorting sounds—nor have they ever been observed singing a song before death. But the idea is suggestive: perhaps animals know when their time has just about come.

As a very old dog, is Ody aware that death lurks just around the corner? Here is what seems to me the most straightforward of the understanding-of-death questions: Do dying animals understand that they are dying, as it actually happens to them? It seems, on first take, something that one could hardly avoid knowing. Yet I am not so sure. I am reminded of a passage from Mark Doty's book *Dog Years*, where he describes the moment at which his dog dies.

> We are talking to Mr. Beau, praising his muzzle and paws and his lovely life, we're holding his face, I'm leaning my head against his belly and praying that he goes easily, trying to send whatever mental force I can muster that might lighten his spirit's way. . . . Each breath enters his chest a little less deeply. And then, when his breathing becomes shallow, he suddenly lifts his head up and back, looking right at me, his eyes widening, with a look not afraid but wondering, startled. A look that would read, were it in a text in a language we knew, as *What's happening to me?* And the life sighs right out of him like a wind, a single breath out and gone.

The end may come as a surprise—I've actually heard many similar reports about the final moments of a dog's or cat's life. Or it may be that some physical change—an agonal breath or final muscle spasm—causes what looks very much from the outside like surprise. Near the very end, the body and mind are both shutting down, so that consciousness of the dying process likely becomes less and less focused—less *aware*—the closer the end draws. It may be, in metaphor and reality, as though simply fading into sleep. Incidentally, I'm not sure that people who are dying are always aware that they are dying, either.

Animals are the perfect students of what spiritual teacher Ram Dass would call "being here now." As I think of Ody aging and drawing closer to dying, this seems right to me. He doesn't *know* that he is different—that he can no longer jump into the back of the car or go romping through the tall grasses after a rabbit. He doesn't draw comparisons to his past. He just *is* different.

And here we come to one of the most troubling questions for me: If I were to ask a vet to plunge that final needle into a vein in Ody's leg, would Ody know what was happening? (And, I also wonder, would he object?) For Ody, at least, my sense is this: he would certainly be aware that things were happening, and because he is a sensitive dog, I think he would know, from my own emotional responses, that something distressing is happening. But I don't think he would be aware that he was on his way out of this world.

What about their *future* death? Ethologist Donald Griffin writes about consciousness of *future* death in *The Question of Animal Awareness*:

> Miller et al., Langer, and others state without qualification that man is the only animal that can be aware of his own future death. But I suggest that we pause and ask just how anyone knows this. What sort of evidence is available either pro or con? Suggestive inferences can be based on the clear demonstration that many social animals recognize each other as individuals, and on the observation that some animal mothers show signs of distress over the corpses of their dead infants, which they carry about for days. How can we judge whether an animal may experience any notion of its own future death after observing the death of companions? The available, negative evidence supports at most an agnostic position.

As with our other death-awareness questions, I think we need to be more specific. How far in the future do we mean? Five seconds? Five minutes? Five hours? Five days? Ultimately, we simply cannot know, but it seems possible that the longer the span of time, the less likely an animal is to be aware of her own impending death. I don't believe that animals live with an awareness of their future mortality, and certainly not with a dread of death. Still, animals undoubtedly live with an instinctive fear of death, and perhaps this constitutes a more robust form of awareness than we think.

As Griffin says, we cannot come at the question of what animals feel about death directly. It is impossible to assess directly the conscious experiences of animals. We must, instead, make do with "suggestive inferences." At the same time, we know much more about animals in general, and about animal consciousness, awareness, mental experiences, and emotions in particular, than we did in the 1970s when Griffin published his groundbreaking book. Scientists are making huge strides in developing tractable neural, behavioral, and physiological measures of consciousness. Were more research to focus on the question of animal awareness of death, we might be able to add still more color to our picture.

DO COMPANION ANIMALS UNDERSTAND HUMAN DEATH?

Several of the most common superstitions about animals reflect a belief that animals can predict human death. A barking dog foretells a death, and a cat (especially a black one) walking across our path means we will soon kick the bucket. Judging from the number of current-day anecdotes about dogs and cats who have stood vigil as their Person died, or had a seizure right before the death of their Person, or generally acted funny right before a death, these beliefs are alive and well. And perhaps there is some truth behind these stories.

The intriguing idea that animals can predict our death has been given a huge boost by a cat named Oscar. Oscar lives at the Steere House Nursing and Rehabilitation Center in Rhode Island. Staff began noticing that Oscar would stake out the rooms of particular patients and would jump on the bed and curl up next to them. These same patients, it turned out, would die within hours. Oscar's behavior was so reliable that staff knew when to call a patient's family and tell them to come so they could be present when their loved one died. Oscar's story gained unusual credibility when Dr. David Dosa, a geriatrician at the Steere House, published an account of Oscar's activities in the *New England Journal of Medicine*.

Although Oscar is the most famous of the death-predicting critters, he is certainly not the only one. For example, Scamp is a miniature schnauzer who lives and works in a nursing home in Canton, Ohio. He reportedly goes into the room of a dying person and paces around and barks. According to an *Animal Planet* report about his activities, he has successfully predicted fifty-eight deaths. Other nursing homes have reported similar death-predicting abilities in a resident dog or cat. Indeed, it now seems to be de rigueur and, a point of pride, for a nursing home to have a death-predicting pet on premises.

A majority of the headlines about Oscar and Scamp include the word "predict" or "forecast," suggesting that supernatural powers are at work. Yet there is a difference between predicting and sensing, and a perfectly natural explanation for Oscar and Scamp's behavior may be close at hand. One theory about how Oscar senses death (if he really does) is that he smells subtle chemical changes in a person's body, such as the breakdown of carbohydrates. This explanation is consistent with what we know about the acute sensitivity of dog and cat noses. Dogs can be trained to detect certain cancers by identifying biochemical markers, can sense drops in blood sugar associated with diabetes, and can give a warning when an epileptic is about to have a seizure. Why not smell when a body is in the early stages of dying?

Some healthy skepticism is in order here but also a generous helping of curiosity. What I find most interesting about these stories is the suggestion that animals might "understand" death in ways that are unavailable and mysterious to us. Animals have incredibly acute senses—much more developed, in the case of smell, than ours. And perhaps, unlike people, animals have an olfactory awareness of death and dying. A little gross, maybe, but pretty cool.

GREYFRIARS BOBBY AND HACHIKO

Greyfriars Bobby was a Skye terrier who lived in Edinburgh during the mid-nineteenth century. So deep was Bobby's love for his Person, John Gray, that he spent fourteen years guarding the grave of his owner, who was buried in Greyfriars Kirk. According to legend, Bobby lay on Gray's grave all day, leaving only to find food. When Bobby died, he could not be buried within the cemetery, since it was consecrated. So he was buried, instead, just inside the gate of Greyfriars Kirk. People were so touched by Bobby's loyalty that a statue was erected in his honor, and it still stands today, welcoming tourists. Even now, his story has such resonance that Bobby has his own website where you can buy Bobby memorabilia, such as tableware, stationary, statuettes, crystal ware, books, and videos.

Hachiko, a beautiful Akita, was taken in as a pup by a professor at the University of Tokyo during the 1920s. Hachiko got into the habit of meeting his owner at the end of each day at the Shibuya train station. They continued this routine until one day Professor Ueno did not return. He had died of a cerebral hemorrhage while at work. Hachiko went to the train station every day for the next nine years, patiently waiting for his owner to come back. As with Greyfriars Bobby, people were touched by Hachiko's behavior, and he attracted a loyal following of commuters who would bring him food and treats (which could be one reason—aside from sheer loyalty to the professor—that he continued coming to the station, day after day). His life is commemorated by a statue at the Shibuya station.

One of my favorite stories is about a bull by the name of Barnaby, who lived in the German town of Roedental. After his owner, Alfred Gruenemeyer, died, Barnaby went to great lengths to reach the cemetery where the farmer was buried. Barnaby jumped over fences and navigated over a mile of terrain, and once in the cemetery he ran straight over to Gruenemeyer's grave. There he stayed for two or three days, refusing all attempts to coax him away. The farmer was considered quite eccentric because he treated his livestock as pets. Veterinarian Klaus Mueller said of Barnaby, "He shows an acute level of intelligence. It seems incredible that a bull could find the exact spot where his master was buried, but he did it."

Many modern-day iterations of these stories are available of dogs, cats, and sundry other creatures who go to great lengths to find their lost loved ones. And—most heartbreaking of all—the tales of animals who (allegedly) die of grief after their human companion disappears. I love these stories, and if I am in a particularly sentimental mood they bring tears to my eyes. But these tales cannot tell us much about whether dogs and bulls really understand human death. Nevertheless, they do tell us something important about animals and human death. These may not be stories about grief or even death, per se, but about attachment, loyalty, intelligence, and sensitivity. And perhaps also stubbornness, force of habit, and in the case of Hachiko, the alluring power of food rewards. It may be that animals have no awareness of death, but they certainly do have an awareness of the loss of companionship, and this speaks to the intense bond that can form between a human being and an animal and that can be equally strong in both directions. Death can feel very much like abandonment, even to intelligent humans. And abandonment is painful. So regardless of what awareness they have, the death of a human companion can be profoundly important to an animal.

BEING SPECIFIC

It is easy to fall into the trap of saying "animals do this" and "animals do that" and to mash together in our minds and our language the vast number of species referred to, colloquially, as animals. But certainly the multiplicity of beings referred to as animals cannot be reduced to any one set of shared characteristics. There is no answer, then, to the question of whether animals are aware of death because this is the wrong question.

Better questions would be, "Are dogs aware of death?" and "Are chimpanzees aware of death?" We can look for species-specific behaviors related to aging, dying, and death. Yet even this is not specific enough. One of the hazards of animal behavior science is the tendency to treat each species as a generic whole. So, you hear such things as "the spotted hyena sleeps during the day and eats at night." But if someone were to say, "human beings sleep at night and eat during the day," we would tell them to stop being absurd. Within each species, different communities (a herd, a group, a troop, a parliament, a murder, and so forth) will have cultural idiosyncrasies. Consider, for example, the bottlenose dolphins in Shark Bay, Australia, who have learned to use sponges as a hunting tool. We also must attend to particularities of age, gender, environmental context, and life experience. As I've tried to show through the delineation of topics in this chapter, our questions can be more specific still, since there many potential points of contact with dying and death: "Do dogs fear death?" "Can dogs smell death?" "Are dogs aware when another dog is dying in their presence?" "Do dogs grieve for other dogs?" "Do dogs grieve for human companions?"

Yet still—if you will forgive me—I'm afraid this is not good enough. As any seasoned ethologist will tell you—and as anyone with more than one dog or cat or bird or rat living in the home will heartily second—even within individuals of a given species there is huge variability in personality, in life experience, in je ne sais quoi. My three dogs are as different as can be, despite their undeniable dogness. In thinking about animal dying and death, we need to remind ourselves always and as much as possible to attend to the individual.

Part of attending to specificity involves what we might call, following moral psychologists, the development of perspective taking. We can try, as it were, to step into the shoes of different animals, to recognize that their point of view will be radically different from ours. Of course the first very difficult step is noticing that animals don't wear shoes and that we need therefore to think way outside the box. But, and this is the key point: we should assume that they *do* think and feel something. Couldn't it be that their "take" on death (each individual and each species) is unique and interesting—and important to them?

The Ody Journal

MARCH 14, 2010

Ody turned fourteen today. We had a big party for him, with my parents and several of Sage's friends. Sage spent many hours in preparation, making cards from all the animals (Ody, Maya, Topaz, and each rat), buying and making presents, and planning the decorations. She made a special cake of canned meat, topped with biscuit "candles." And I made Ody his favorite dinner: homemade hamburger, rice, and shredded carrot.

I think this will be his last birthday. But that is what I thought last year. Ody is inscrutable.

MARCH 18, 2010

Ody is thirsty a lot and drinks often. Usually after a big drink (often out of the toilet) he vomits it up on the floor. This is probably why he's always thirsty, this and the fact that he's always panting, which must make him dehydrated.

MARCH 27, 2010

Yesterday I was determined to help Ody work through whatever psychological barrier prevents him from coming in the dog door, this after having to get up three times the previous night to let him in. I've seen him come in on his own—saw him do it yesterday—and I know he can do it, physically. It's a mental barrier that's the trouble.

So, next time I notice Ody going *out* the dog door, I arm myself with a slice of his favorite American cheese and wait inside by the opening. When he barks to come in, I lift the grey flap of plastic and call him, holding out a treat so that the smell tickles his nose. He hesitates, I move the cheese a bit closer, and then he gingerly sets one foot on the little step stool and sticks his head through the door. I wave the treat again. Now he tries in earnest.

Three legs make it through and one doesn't. Ody is pretzeled—one hind leg sticks straight out the door, the other collapses, and his front paws scrabble on the slick bamboo floor covering. The scene reminds me of Winnie the Pooh stuck halfway through Rabbit's hole, except this isn't funny. Ody simply doesn't have the mobility to extricate himself, and he flails and struggles. His eyes are cloudy pools of trouble.

I try to help him from the front side. I think that if I lift his body, I'll be able to pull his hind end through. But he tries to bite me. So I head out to the porch and push him from the back side, helping the recalcitrant leg up and through. He is free now, but the hind leg is cramping and stays splayed out stiffly. For a while he can't move.

The next day, it dawns on me. We need to build Ody a big wide ramp up to the dog door. Maybe then he would be okay. It's the step up to the door that causes him trouble. I explain to my spousal carpenter what I have in mind, and he sets to work.

MARCH 30, 2010

I made Ody go on a hike yesterday with Sage, her friend Annalynn, and me. It was cold out, and Ody always does best in the cold, and I figure it is good to keep him active. We had planned to hike the Coulson Gulch trail up in Big Elk Meadows, but the road was too snowy and we couldn't get up to the trailhead. So instead, we parked at the base of a four-wheel-drive road that has always intrigued me. Annalynn was excited to let the dogs out of the back of the car and ran around and opened the hatch before I had enough wits about me to call a warning. Out leapt Topaz, pushing past the others. Out leapt Maya, already hot on the scent of a squirrel. And out leapt Ody, as if he believed he could fly. His front legs crumpled on impact, and his body bore down, nose first, then head, then the rest of him. He rolled a couple of times and came to a stop in a contorted upside-down jumble of legs. Once we righted him, he shook once and stumbled up the trail as if saying, "I'm fine, really. It looked as though it hurt, but it didn't. Really."

I love this about Ody. Time and again, he maintains a strong faith in his ability to fly.

The snow was deep—much too deep for Ody. But he seemed to enjoy himself. He shuffled along, doing his heavy pant (despite the cold), stopping often to sniff and pee. He doesn't *amble*, as you think an old person or dog might; rather, he walks with intention, as if he is on a mission. What does he fancy himself doing?

When we got home, Ody slept more deeply than usual. I got worried because he didn't eat dinner and even refused a piece of liver treat. But the next morning he seemed chipper and alert.

Mark Doty, on dogs' relationship to time: "Because dogs do not live as long as we do, they seem to travel a faster curve than human beings, flaring into being and then fading away as we watch."

I told Sage, when she was little, that Ody's hairs were seeds, and that if we picked one and let it be blown away by the wind, a little Ody would grow where the seed landed. One of her chief entertainments during the long drives between Colorado and Nebraska was planting Ody seeds. Sage used to pick a fingerful and let the red hairs be pulled into the wind by an open car window. Or, like the airy white seeds of a dandelion, to be blown away in one big breath. Breath as life. We planted little Odys all over the Midwestern plains.

Limen: a threshold, the point at which a stimulus is of sufficient intensity to generate a response. "Such to the dead might appear the world of the living—charged with information, with meaning, yet somehow always just, terribly, beyond that fateful limen where any lamp of comprehension might beam forth."–Thomas Pynchon, *Against the Day*

Ody's new nickname: PegLeg. He walks like a pirate with wooden legs; they don't seem to bend anymore.

MARCH 31, 2010

Ody gets stuck if I drop his leash to let him roam at the park. He steps on the leash with one of his back feet, and then just stands there, as if a phantom is holding him back.

APRIL 1, 2010

Watching Ody stand up from lying is like watching a kite trying to get airborne when there is not quite enough wind. Too much weight drags down the back while the front strains to catch a whispered current of air. Not enough lift; too much gravity. Ody's back end keeps him from soaring up alongside the birds. Sometimes I see him simply give up and stay where he is.

Ody used to be a prized biscuit catcher. No matter how far, how high, how incompetent my throw, somehow his mouth found its way around a treat, clamping down with a crunch. I no longer toss Ody biscuits. His eyes are too dim to see the flying projectiles and they simply hit him in the face. When a biscuit lands even a few inches too far for him to reach, he has to make a monumental effort to get himself up and to the biscuit. It is a race against time. Will Topaz reach it first?

APRIL 2

Chris built a sturdy wooden ramp leading up to the dog door, to replace the little step stool. A handicapped ramp into the house. Ody refuses.

APRIL 3, 2010

Ody has had a noticeable limp these last two days. He has seemed eager to go for walks, but has a hard time walking. Ody's thirteenth year was the Year of No Walking. He went through a long phase where he just didn't want to leave the house. Now he once again likes to go, most of the time. I still have to do some serious work to rouse him from sleep, and coming out the door is still a psychological battle for him. But once we're in the yard and the leash is on, all is well (assuming, of course, that the sky is free of those fluffy white portents of danger).

He still won't use the wooden ramp to the dog door, even though I've tried coaxing him with hotdogs.

APRIL 4, 2010

Got Ody a doggie waterbed today, on a friend's recommendation. He said his elderly dog loved his waterbed, and that it helped his sore joints. So far, Ody refuses to lie on it.

APRIL 6, 2010

Ody fell off the couch today. He was sleeping with his back facing out, legs extended toward the couch cushion. But he was a little too close to the edge. His body just slid right down with a soft thud onto the floor.

He also got trapped under the trampoline. We couldn't figure out where he was, and looked all over the house and yard and up the street. Finally heard scuffling noises and found him under the tramp, unable to find his way out. His days are full of misadventures.

APRIL 7, 2010

Ody just tried to drink from the fish pond and fell in. I had a heck of a time fishing him out.

APRIL 10, 2010

Ody's tail is an emoticon, shorthand for his general feeling toward the world. Curved downward like an upside-down smiley face means unhappy. Straight outward means all's well. Upright and going back and forth is a sign of intense interest or excitement. Lately, his tail almost always curves down, except on rare occasions, as when Ody meets Finn (the neighbor's dog). Not sure if this is a physical change (do tail muscles atrophy?), or an emotional shift.

Maybe he has degenerative arthritis of the tail from the time he injured it trying to chase a squirrel up a tree. He was jumping up on the tree's trunk, trying to reach high into the branches, and somehow he ended up coming down wrong and landing straight on his tail. There was nothing the vet could do for a sprained tail, and Ody had to suffer the indignity of living with a bent tail for a month or two. During the bent-tail episode, Ody was not at all himself. I wondered if it was because Ody couldn't smile with his tail. You know the trick of forcing yourself to smile, and this then actually improving your mood, because the facial muscles stimulate the brain to release chemicals that make us feel happy? Ody had to frown for several weeks straight.

APRIL 20, 2010

Got Ody a new soft bed for under the piano, since the other was sort of lumpy. Oatmeal colored, with black dog-bone shapes. He likes it!

APRIL 21, 2010

We have to shut the dog door at night now because Ody's barking outside wakes the neighbors. Every night now I wake once or twice to Ody's barking to go out. Usually he rings at about four; this seems to be his witching hour. I always wake with a little rush of anxiety because I think maybe he has gotten outside and is disturbing the neighbors. What if we forgot to cover the dog door? The nights are a total blur of dog activity. Ody phantom barking, Ody really barking, getting up to let Ody outside, waiting on the couch to hear him click-click back inside so I can shut the door and stumble back to bed. Maya tapping her paw on the bedside so I can tell her it's okay to get up; Topaz doing the same. Many days I feel like a zombie. I am so tired by evening that I often crawl into bed at 8:30, apologizing to my husband and daughter for checking out so early, begging indulgence.

There are periods of a few days when Ody will sleep through the night. Still, though, I hear him. I wake to the phantom barks. A bit of adrenaline titrates into my bloodstream. I am anxious about the neighbors, anxious that Ody is desperate to go out and relieve himself. I stumble out to the glass doors, looking for Ody's shadow there in the dark, waiting to be released. But he is not there. Wasn't it he that barked? I walk into the living room and feel along the length of the couch, looking for his warmth. Not there. I peer under the piano, at the mess of dog waterbeds and dog-bone cushions, and there he is, out cold. He doesn't wake unless I put a hand out and touch his side—but I don't do this because it always startles him. I suspect that I will hear phantom barks a long time after Ody is gone.

Ody seems to have a little extra spring in his step when I walk him alone. I don't do this often because of the rueful looks from Topaz and Maya when I shut the door in their faces. Maybe Ody has more fun by himself because he doesn't suffer the humiliation of being last the whole way, of being dragged along barely able to get his peg legs moving fast enough. I know I'm anthropomorphizing here. I'm not sure that being unable to keep up with the other dogs would even cross Ody's mind. Can't help it.

Things he most enjoys on the walk: peeing, especially re-marking over Maya's scent; sniffing where other dogs have been; saying hi to children at the park.

Ody has a look of worry about him. He seems especially apprehensive on our walks. Maybe, like a frail elderly person, he is concerned about the basic mechanics of getting around. And he does scare me. He walks not in the center of the sidewalk, but over at the edge, right by the curb. On our street, this isn't too worrisome because the sidewalk meets the street in a low, gentle curb, only about two inches high. On the larger street out front, the curbs are higher. Ody usually manages to slip off at least once per outing. I stay behind him, on pins and needles, shadowing him with arms ready to reach out and catch him.

Dogs fall into different philosophical schools. Some dogs are rationalists; a great many are hedonists and Epicureans. Ody is without doubt an existentialist, through and through. Kierkegaard reincarnated, I think. Dread on a grand scale, woven into every fiber of his being.

I'm in New York City with my brother for a few days, missing the dogs, yes, but relishing the full nights of sleep and the break from worrying. We're in a Greenwich Village café, just ordering breakfast, when Chris calls. I step outside into the rising heat. Chris unleashes a torrent, his voice loud and clipped. It is garbled, but the topic is clearly Ody, the dog door, barking in the middle of the night, and neighbors. What I finally determine is that Chris had forgotten to close the dog door before bed and Ody had gone outside in the middle of the night and woken the neighbors again. This is after we had solemnly promised our very patient neighbors that it would never happen again.

Then Chris drops the bomb. "If Ody gets out and wakes the neighbors one more time, we're going to have to put him down. I'm serious."

I stiffen at these words, even though I know they aren't true. I would never allow this, and Chris would never go there, really. And if it came down to a battle, I would win. I would have the moral upper hand, no question about it. The first wave of Chris's venting always tips to the extreme, and then mellows. Ody has been threatened with the needle before. (Origin of the "blue needle" threat: a friend who worked as a veterinary technician told me years ago that animals were euthanized with a colored syringe, usually pink or blue, to distinguish it from "medicinal" solutions. It became a not-so-funny joke in our household, when Ody did something outrageously naughty. "Ody, do that one more time and you'll get the blue needle." Of course we never meant it.)

All day long the conversation hovers around me, having a slight dampening effect. I feel annoyed with Chris for blaming Ody—it was Chris, after all, who left the dog door open not once, but twice, so that Ody woke the neighbors. But ultimately, I always feel responsible. I am the first line of defense for the dogs and take the heavy artillery. The dogs are "mine" more than anyone else's. Look for the dogs in the house, and they will be with me. I feed them and walk them and train them and take them to the vet and know what time they eat and what they like and that greedy Topaz is actually the pickiest eater of the bunch and that Maya gets cold at night if she isn't under the covers and doesn't like to get her feet muddy and which kind of cheese Ody likes best (squeeze cheese, American style!).

It was my idea, each time, to adopt one dog and then the next. I begged and pleaded for Ody; convinced my husband that a puppy was absolutely necessary for then five-year-old Sage (Hello, Maya); and then launched a protracted campaign, five years later, for Sage to REALLY have her own puppy, which she was now old enough to raise and train (Hello, Topaz). Looking at this chronology, I see a five-year pattern. Which means that since we've had Topaz for about two and half years now, in another 2.5 years I need to paste a very large post-it note on the refrigerator: NO MORE DOGS! RESIST THE CALL!

At the end of the conversation, Chris announces, "I'm going to screw the dog door shut permanently."

"Look," I say in my most even, Obamaesque tone, "I am really sorry that you're stressed. I know it is a lot to handle." But please, I want to say, please don't close the dog door forever. The ability to go in and out when they want is one of the dogs' few shreds of freedom. Please don't be mad at Ody. Now is where I am supposed to reframe the problem, come to a cooperative, creative, and negotiated settlement that seems to offer a positive path forward. But I find myself at a loss for words and at a loss for solutions.

MAY 9, 2010

When I read books about life with dogs, like *Amazing Gracie* and *Merle's Door* and *Dog Years*, I feel so envious. Although these dogs have foibles, they seem to add nothing but sunshine and love to their humans' lives. And although my dogs add more sunshine and love than I could fit inside the state of Texas, they also add a sizable hunk of stress and pain.

MAY 10, 2010

I can't get Ody to go outside before bed without making a Hansel and Gretel–like trail of hotdog. With the first chunk, I stir him from sleep. I hold the hotdog in front of his nose for a few long moments, until his nose twitches and his eyes slowly crack open. Chunks two and three rouse him off the couch. A fourth leads him into the dining room. The kitchen—the Topaz war zone—requires two or three chunks, one every couple of feet. Ody gobbles and then looks up at me with those cloudy eyes, assessing what? A few more chunks and he is at the threshold. But he still won't go out, not without more chunks. All this time, Maya and Topaz hover around, expecting their fair share, one each, one two three, never in the same order (no playing favorites!). I throw a chunk far out into the yard, so that Topaz will run after it and Ody can feel safe enough to inch forward.

It takes about ten minutes to lure Ody outside; then another long time to coax him back in. The return trip is without hotdogs. Just me standing by his side, offering words of encouragement and keeping Topaz at bay.

All this just to get Ody to pee, with the self-interested motivation of buying myself a few more hours of sleep.

MAY 11, 2010

The dogs have different loci of expression.

For Ody: the tail (curved up or down) and forehead (crumpled, smooth, lifted, relaxed)

For Maya: the eyes (wild eyes, eyes of love, hard eyes protecting food)

For Topaz: the ears, definitely the ears. Bigger than any dog ears I've ever seen. The ears, when he's listening to something, point up and straight forward, shifting ever so slightly to and fro, like twin submarine periscopes. Scanning the horizon for enemy warships or maybe pirates.

MAY 16, 2010

Ody mostly lies in his place under the piano. Alone. He lies there watching life pass him by—dogs and people, coming and going. Does he isolate himself because this is what he wants? Does it please him to lie under there? Or is he trapped by fear of Topaz?

I think he is lonely, but perhaps I am the one. From my hideout, I watch the world pass by.

MAY 17, 2010

Ody throws up a lot these days. Mostly it is just water mixed with slobber, but sometimes it is yellowish; occasionally he throws up a huge pile of undigested dog food (kibble still intact, proving that he does, indeed, vacuum his food rather than chew it).

MAY 19, 2010

Talked to Pansy about leaving Ody when we go to Gunnison. She's fine with it, but suggested leaving Maya, too. Maya is mother hen to Ody when they are at Pansy's. Maya watches after Ody, checks on him throughout the day, and sleeps next to him, curled up with her head resting on his back. Pansy said that last time the dogs stayed, Ody couldn't go down to the basement because he can no longer navigate the stairs. Maya and Topaz were down there, and she put Ody in a new place, in the little carriage house. Ody cried all night because he was separated from Maya.

MAY 21, 2010

Thinking about vacations differently now. We had planned an eight-to-nine-day roam, beginning in Gunnison with the Sage Burner race, then moving on with mountain bikes to Crested Butte, Moab, or wherever struck our fancy. I didn't think about the old dog details until close up and now I'm realizing that it just won't work with Ody. He will have a hard time being in the car so much and won't be able to go to hikes or bike rides, and it will be far too hot for him to stay in the car. We've changed our plans and will now spend just three days in Gunnison, at a motel, and then head back toward home and spend the remainder of our time at the cabin in Estes Park. Ody will enjoy this.

Ody lay in the sun on the porch yesterday and I noticed something I've never seen before: along the ridge of his ribs and curving along his body is a line of white hair. In most light it is unnoticeable. Or perhaps I have failed to look closely enough. He is inscribed by age.

MAY 25, 2010. GUNNISON.

Glad that we brought Ody. He has really perked up and seems to be having fun. He wanders all around the motel's grassy courtyard. It is funny to watch: the other dogs dart around from one spot to another, with their doggy ADHD. Ody plods, but he does so with deep concentration and intention. He doesn't go toward other dogs, birds, squirrels. What is he after?

First night in the motel. Ody began to bark about 1:00 a.m. Chris got up and let him out into the courtyard. Ody started to wander off. Chris, in bare feet and underwear, whisper-called "Ody!" Ody kept going. Chris caught up with him and tried to catch his collar. Ody reached around with toothless muzzle and tried to bite. Chris ran back inside to find a headlamp and some shoes and went back out to find Ody, who in the meantime had been steadily plodding away. When Chris finally found Ody and tried to grab his collar, Ody reached around, again, to inflict a toothless bite. Chris had to go back inside, once again, to find a leash to lasso Ody and pull him in.

Ody continued to bark in the motel room, even after his excursion into the courtyard. Finally, Chris put Ody in the car. We could still hear the muffled barking. Our motel neighbors must hate us. Will the manager ask us to leave?

All morning Ody's teeth chattered (it's cold here). After it warmed up a bit, Ody was chattering and panting at the same time.

Today we got Ody a bear bell so he can wander freely around the motel courtyard but we can still locate him.

JUNE 1, 2010

Since her birth, Sage has considered Ody her sibling. The fact that they are not of the same species didn't seem to occur to her. Their close bond was evident from the beginning, eleven long years ago.

Sage's first word: "dog"

Sage's first full sentence: "Bad Ody eat dirty diaper."

JUNE 3, 2010. ESTES PARK.

My worries came true today, and when I had my back turned, too. Ody fell down the cabin stairs. We had come back from town and let the dogs out of the car. Ody seemed to want to stay down in the driveway, so I went up the two flights of stairs to open the door and drop off books and sweaters. Chris and Sage went up, too.

As I was unlocking the door I heard Sage shriek "Ody!" I dropped the sweaters and books and ran back across the porch and there was Ody in a crumpled heap on the lower landing, covered with dirt and pine needles.

Chris carried him up the stairs and deposited him gently on the porch. Everything looked intact and he hobbled inside. I think he's okay, but I notice a couple of small spots of blood on the sheet where he's been sleeping. I can't find the source of the bleeding.

JUNE 4, 2010

Ody fell in the creek up the road from the cabin. He went down for a drink, looked carefully at the topography, seeming to assess his ability to navigate down the bank. He must have decided it was doable; he put one foot in the water, then the other. Everything was okay. One step more, right, left, and then it all goes to hell. His front legs sink way down, and his face dips under the water, rear end still in the air, his tail like a snorkel. I see panic spread through his body, and he begins to flail. He twists around to face the bank and manages to position his front paws up on land. Then his tail begins to helicopter, as if he thinks it will propel him out of the water. Chris, who is closest, runs over and pulls him out of the water. Ody shakes, almost toppling over in the process, and shuffles off up the hill to the dirt road. Then, as if exhilarated by his brush with disaster, he launches himself over a deep trough in the road. I cringe and hold my breath, but somehow he lands it. And off he goes down the road at a rare brisk trot.

3

Old

The changes are so gradual that I almost don't see them happening. It is like watching my daughter Sage grow up. She changes a little bit every day, grows a millionth of an inch, and becomes a fraction more adult. I see, yet don't see, these changes. And then all of a sudden, I will be overwhelmed. Like sitting at her sixth-grade end-of-the-year slide show, watching images flash past of all these children—many of whom I have watched grow since kindergarten. One is laughing with a friend outside the lunch room; another sits engrossed in a science project, chin resting on fist. They look so serious. So big. I have that tight feeling in my throat and my eyes tear up. *They are no longer children*, I think. *How did this happen without my seeing it?*

Change has been like this with Ody—bit by bit, right under my nose, but beyond my view. At about age eight, the lumps began slowly growing on his belly and sides, the white hairs began to overtake the red ones on his muzzle, and the skin tags began to appear one by one. I noticed, but didn't really take it in. It was too incremental, and I was too close to really see. I would have an occasional epiphany, when I would think to myself, *My God, he's really getting old*. Now, at thirteen years and eight months, it is no longer so gradual.

Perhaps the most obvious sign of Ody's age is the state of his teeth—and from what I gather, the state of one's teeth is a pretty good indicator of age, particularly for animals without a good dental plan. Some species, such as mongoose, goat, shrew, and African elephant, can even die of old teeth: nature gives them one full set and when these wear out, their odds of survival dwindle. If Ody were a wild dog, he would be in serious trouble. From the age of about thirteen on, Ody has had exactly one-half of a canine, and this stub is brown and cracked. Every time I'm in the vet's office with Ody, I ask them, "Do you think his teeth hurt him? Does it hurt him to eat?" They assure me that he probably isn't too bothered by his lack of teeth, and eating, certainly, is not a problem. It amazes me what he can eat with one-half of a tooth.

Before you blame his dental situation on our neglect, consider some of the larger items Ody has eaten during his life: three doorframes, three couches, two mattresses, a 4 x 4 piece of wood, several plastic indoor kennels with metal doors, the metal frame on an outdoor chicken-wire kennel, innumerable bedspreads and chair cushions, and the entire backseat of a Subaru Outback.

Ody doesn't seem to care that he has brown stubs for teeth. Nor does he seem concerned with many of the other age-related changes in his appearance that might bother a vain or feckless dog. He doesn't mind that his fine red coat is now mottled with white, or that his well-muscled legs have withered into thin stalks. Nor does he seem to notice the fleshy barnacles that hang off his face and body, the creepy black growth that dangles off his catfishes or the dark goo that accumulates on his lips and rubs off on the couch when he's sleeping.

Not only has his appearance changed, but so has his behavior. When we go for walks, he pees on everything. If you aren't paying attention, he'll pee on your shoe. His legs are so unsteady that he rarely lifts a leg anymore. Instead, he squats a little, halfway to girl dog position. Often, he'll begin walking before his brain gets the signal to stop urinating and he'll leave a long dribble-line up the sidewalk. He poops as he walks, too, leaving lumps on the sidewalk, like a trail of chocolate chips. These are too small and squishy to pick up effectively, so the sidewalks in our neighborhood are littered with brown spots.

I worry about Ody getting lost because he has a penchant for wanderlust. If I let him out in the front yard, he'll just amble up the street and I have to run after him, calling to the neighbors as I pass, "Has Ody gone by?" Since he can't hear, I have to rely on sight to find him again. Once he got lost in our backyard, having for some unfathomable reason decided to crawl into the space under the trampoline. He couldn't figure how to find his way out, so he just stood there panting until finally, while reconnoitering after a search of the neighborhood, we happened to hear his heavy breathing coming from the back side of the house.

He pants constantly. He stands in the middle of the kitchen while I cook, and pants. He stands next to the table while we're eating, and pants. His breath is terrible, and easily travels the length of the table. When we walk around the block, he pants, even if it is 20 degrees and snowing. The vet explained to me that Ody suffers from laryngeal paralysis, which is not uncommon in older dogs, especially large breeds. The laryngeal cartilages that open and close the back of the throat don't function properly, so air can't move in and out as easily. As sick and tired as I get of the panting, I can't be angry at Ody; he can't help it.

The paralysis also causes vomiting. Every time Ody drinks, he makes a horrible retching sound, followed by dry heaves, and then the water comes hurtling back up his throat and onto the floor. Which, of course, leaves him thirsty, so he drinks again. There are little pools of spittle-flecked water all over the house, especially in the hall outside the bathroom where his water bowl is located. Put on fresh socks and you will most certainly step in one of his puddles.

Ody often chatters his stubs of teeth. I don't know why. Sometimes he chatters when he's cold, which makes sense. But he also chatters when he's hot, and sometimes just for the heck of it. Sometimes he pants and chatters at the same time.

And, the pièce de résistance: the bark. Young Ody was never been much of a barker, but when he did vocalize, it was a deep and sonorous and, I have to say, a very beautiful baritone. Now, Ody's voice has transformed into a raspy croak. When I described to the vet how Ody had started sounding like a canine Darth Vader, the vet immediately said, "Laryngeal paralysis. The bark is unmistakable." Even more noticeable than the change in sound is the new rhythm. Ody doesn't release a sudden volley of barks and then stop, as he used to. Instead, he croaks just once, waits for ten or twenty seconds, and croaks again just once, and so on, like a fire alarm beeping its low-battery warning. The lapse between each bark is just enough for you to almost gain your composure, and then it comes again. The bark will continue for as long as it must; there is no waiting him out. If he is inside, he barks to go out. And if outside, in. Sometimes he gets trapped in one room by Topaz and will bark until I rescue him. Oftentimes he just barks for no reason that I can discern. Nights are now routinely interrupted by this unusual chorus.

I am surprised by aging. It never occurred to me when we took home a fat wiggling Ody that dealing with the end of his life would be so hard. I thought about the puppy stage, which is difficult but oh-so-fun, the adolescent stage, which is challenging but brief, and then adulthood. Maybe, if I thought about old age, it was with a small sigh of relief: old dogs just sleep all the time. They are content just hanging out at your feet and don't always need to be walked or played with. Low maintenance. How wrong I was. Having an old dog—especially having an old Ody—is a lot of work and requires huge amounts of patience and special kindness. And, perhaps more important, it requires constant adaptation to his changing needs.

Aging can be hard on animals and on their human companions. But the challenges of aging can invite us to know and love new dimensions of our animals, as we become particularly attuned to their evolving needs. It is a time for us to give back some of the unconditional love, patience, and tolerance that our pets offer us throughout their lives.

BIOLOGY OF AGING

Scientists understand a good deal about the process of aging, but there is still much that we don't understand. And no one, yet, has developed a cure.

Lest you think that aging is straightforward, consider all the ways in which you can age: chronologically (how many years old you are); societally (whether the society in which you live considers you "old" and how you are expected to behave at a particular age); biologically (the physical state of you as an organism); functionally (two people of the same age can differ markedly in their physical and mental capacities). A population can age, too, if there is an increase in number and proportion of aged people. Aging in human beings, then, is considered a multidimensional process, with physical, psychological, and social elements. Aging in animals is similarly multidimensional, and this will have important implications for living with elderly companion animals. Unfortunately, the literature on psychological and social aspects of aging in animals is still relatively small. When scientists talk about aging in animals, they generally mean biological aging.

Senescence is the state or process of biological aging and the changes that take place in an organism as it ages. Senescence occurs both at the cellular and at the organismal level. Cellular senescence refers to the aging of individual cells: normal diploid cells lose their capacity to divide, usually after about fifty cell divisions, at least in vitro. Red blood cells, for example, live for about 120 days within the human body. Organismal senescence is the aging of an entire organism. According to my reading, there is general agreement that organismal aging is characterized by reduced ability to respond to stress, increased homeostatic imbalances, and increased susceptibility to disease. Death is the ultimate end point of aging, though it is not age, per se, that kills us. Old age is not a scientifically recognized cause of death.

It comes as some surprise to me that there is no accepted theory of biological aging. Some think aging is programmed into genes; some think it is the accumulative damage of biological processes; and others, some combination of these. We don't need to go into detail on the science—it gets very complicated very fast—but simply observe the vibrancy of aging research. (Note that theories about why organisms age are distinct from theories about why organisms die: despite obvious connections these are biologically distinct processes.)

VARIETIES OF AGING

Being old just means something has been around for a long time. Mountain ranges are old. To be aged means that certain biological processes have taken place.

I was surprised to learn that aging is hard to define. According to neurobiologists André Klarsfeld and Frédéric Revah in *The Biology of Death*, aging can best be defined statistically, as an increase in mortality rate with age. In other words, aging means you run a greater risk of dying with each passing day. Although it has a ring of scientific credibility, this definition makes me a little anxious: tomorrow will be slightly more dangerous than today. Aging is clearly tied to death, but Klarsfeld and Revah reassure us that aging never killed anyone: "Aging is not so much a direct cause of mortality as the ensemble of processes that increase the vulnerability of the organism to the direct causes, such as infections, tumors, or occluded or broken blood vessels."

Scientists have found tremendous variety in how living organisms grow old and die. Human beings, for instance, may live up to 120 years, while mayflies live for several minutes. Three basic strategies of aging have evolved: acute senescence, gradual senescence, and negligible senescence. Some species age all at once and die in short order. Certain bamboo species live for seven, thirty, sixty, or 120 years, then bloom and die. Sudden death sometimes follows right on the tail of reproduction. In certain species of spider, for example, the female devours the male right after or even during copulation. "Precipitous decrepitude," according to Klarsfeld and Revah, is used to describe what happens to salmon, who go back to their birth home to reproduce and then immediately die. As mammals, we are all familiar with gradual aging, which is just what it sounds like—growing old gradually over some species-specific span of time (whether it be 120 years or only two or three). Some species, such as the bristlecone pine, the sturgeon, and the quahog, show "negligible senescence." The hydra (a simple freshwater animal) does not undergo senescence at all, nor does the *Turritopsis nutricula*, a species of small jellyfish. This doesn't mean they don't die. They eventually do, of disease or trauma. But they are, as scientists say, "biologically immortal."

Lifespan is a measure of how long you would live if all possible environmental hazards were removed. A species' lifespan does not indicate how an organism ages, since a species with acute senescence can have the same lifespan as a species with gradual aging (e.g., bamboo and human being). Each species has a normal lifespan, determined by genetic makeup, physiology, and evolution. And, in fact, within species we can find considerable variation. For example, the average lifespan of a Vizsla is 12.5 years. A miniature poodle might survive to almost fifteen, while an Irish wolfhound will be lucky to make it much past six.

What is the longest-lived creature on earth? Certain blueberry bushes have lived for thirteen thousand years, and there are bacterial spores that are twenty-five million years old. A Great Basin bristlecone pine, named Methuselah, was purported to be 4,789 years old when it was sampled in 1957. Some speak in hushed voices of a creosote plant that is 11,700 years old. Among mammals, the prize for maximum lifespan may (though there is some controversy) go to a bowhead whale that is at least 211 years old.

WHAT DO WE KNOW ABOUT OLD ANIMALS IN THE WILD?

Biologists and ethologists categorize animals based on their age, recognizing that each life stage is physiologically and behaviorally unique. But there is no official, scientifically recognized category for the aged, even though animals continue to go through distinct physical and behavioral changes as they move beyond adulthood. Anne Innis Dagg's 2009 book, *The Social Be-*

bavior of Older Animals, is the first and only full-length monograph to focus on the behavior of elderly animals. Is there ageism in animal science too? Or are there simply no old wild animals to study? Dagg suggests an answer at the beginning of her book: "Until recently, people believed that wild animals did not live to be old, dying instead from accidents or disease, or being killed and eaten by predators." But this assumption, it turns out, is wrong: surprisingly many animals do survive into old age. It is high time to study this ghost population.

Since there isn't a body of literature specifically focused on aged animals, I decided to look for data on aging in the myriad books devoted to particular species. I looked through all the books in my local library on wolves, elephants, whales, dolphins, and orangutans. And what I found was fascinating, in a frustrating kind of way: the indexes have no entries for "age," "aged," "elder," "elderly," "old," or "senescence." I couldn't think of any other ways to say "old," so I looked for entries on "death," "dying," "mortality," "lifespan." This yielded a bit of fruit, but not much.

From the very thin data on wild animals, here are a few things that we do know. The chance that an animal will die in extreme old age is relatively slim (this is true for people, too). For mammals, at least, the first year of age is the most dangerous. And the older an animal gets, the greater its chance of dying, statistically speaking. Still, old is a relative concept. Wolves in the wild may be quite old at age six—life wears them out fast—even though wolves in captivity can live as long as twenty years. Incidentally, some species die earlier in captivity. African elephants, for instance, live well into their fifties in the wild, but median life expectancy in captivity is only seventeen years.

Old animals move less quickly and are less agile than younger animals. Many animals display obvious physical signs of age: they become gaunt and their fur or hair may turn gray or white and may become increasingly thin. (There are exceptions: the plumage of old swallows is indistinguishable from the plumage of younger ones.) Males and females usually have different life spans, with females holding a slight survival advantage over males. Old animals are more likely to suffer from disease than are younger animals. Aging decreases the ability of prey to escape predators and, conversely, can decrease predatory performance (a phenomenon known as "predatory senescence"). The elderly are unlikely to be dominant in their societies, though they often play important social roles. For example, a recent study by Karen McComb and colleagues suggests that, among African elephants, the groups led by older matriarchs are more successful at fending off male lions than groups led by younger females.

Although sick and injured animals are much more vulnerable to predation, these conditions don't always spell immediate death. Animals of some species will care for one another, and the sick or aged will sometimes be given extra protection. Consider, for example, Darwin's blind pelican, who was cared for by his flock mates; primatologist Robert Sapolsky's observation of baboons taking special care of a group member who suffered from palsies and seizures; and the well-deserved reputation of elephants for ministering to ailing herd members.

OLD PETS

In contrast to old animals in the wild, who have attracted relatively little attention, aging pets are a subject of considerable interest. Within the population of companion animals, the elderly is the fastest growing category with over 35 percent of all pets in the United States now considered, by their vets, to be geriatric. There are about seventy-eight million companion dogs in US households and ninety-four million cats, which means roughly twenty-seven million geriatric dogs and thirty-three million geriatric cats. These numbers are likely to grow, as veterinary medicine offers an ever-wider range of treatments, from organ transplants to hip replacements, and as better lifelong care increases pet life expectancies. In step with the changing pet demographic is a growing appreciation for the final stages of our companion animals' lives: there are geriatric specialists, old-dog and old-cat foods, products designed help older animals maintain functionality, books devoted to caring for old pets, advice from trainers about how to deal with age-related behavioral changes, and old-dog and old-cat rescue organizations.

Despite increasing attention to the needs of old companion animals, for many of them, being old is a dark and unpleasant stage of life. There remains a deep prejudice against the old. Sometimes dogs and cats are euthanized merely because they are old, even though they may be reasonably healthy or have treatable problems. Many more languish in shelters, where adoption rates for seniors are very low. Old animals too often suffer from untreated disease and pain, either because owners don't recognize their changing needs or because they cannot or will not pay for adequate veterinary care.

Ironically, at the same time as the population of pets in American and other wealthy countries is undergoing a demographic shift, with many more animals living well to old age, in the wild, things may be moving in the opposite direction. The rate of dying in the wild is increasing for many species—and life expectancy is dropping—sometimes dramatically, as climate change, habitat destruction, and pollution (including oil spills) challenge survival capacities. Polar bears, for example, are much less likely to reach old age now and are less likely to survive their first year, as compared to even a decade ago.

WHAT TO EXPECT WHEN YOUR PET IS SENESCING

According to the veterinary literature, dogs and cats are considered geriatric when they turn seven (five for some larger breeds of dog, nine for some smaller breeds). My little Maya turned seven yesterday, which means she is now, officially, a senior citizen. She is still active, and when she's out running with me, she looks youthful and beautiful and svelte. But I notice that she sleeps a lot these days—and she is getting those lumps, just like Ody. Only lipomas, says the vet. Maya also has skin tags on her eyebrows and chin, and the fur beneath her eyes is streaked with white. In dog years, Maya is just about my age—midforties—and Ody is in his late seventies.

Aging brings with it many physical changes for dogs: skin and hair disorders and alterations, including the distinctive graying of the muzzle; reproductive system changes, especially for male dogs (male dogs who have not been neutered often develop prostate problems; there is no canine menopause); bone and joint problems, including osteoarthritis; muscular atrophy; decrease in heart, lung, liver, and kidney function; intestinal issues (constipation, gastritis); weakened immune system; vision changes; and hearing loss. In dogs, smell is usually the last sense to fail.

Aging also affects the brain. For humans, cognitive decline is said to begin in our thirties—a fact I find totally depressing, since I'm clearly headed south. The aging brain undergoes structural and chemical changes, such as loss of neural circuits and brain plasticity, thinning of the cortex, and declines in dopamine and seratonin levels. We can't think as fast, can't remember as well, and can't process as much information. Cognitive decline also affects animals, sometimes quite dramatically. The vet tells me that some of Ody's behavioral mannerisms, like his knuckling, are caused by neurological decline. Ody also probably has some level of canine dementia, or CDS (cognitive dysfunction syndrome). The brains of dogs with senile dementia are more or less identical to those seen in cases of human Alzheimer's disease. Necropsies performed on dogs with CDS show a similar kind of plaque to that found in Alzheimer's patients; because of the similarities, dogs are being used as models in Alzheimer's research. Like Alzheimer's, CDS is not curable, but drug treatments (e.g., Anipryl) have been shown to help slow a dog's decline and can sometimes decrease symptoms.

Changes in the body and brain can in turn affect behavior. Dogs with mental deterioration may stop being as interested in their owners, sleep more, become incontinent or disoriented, and show changes in personality. Sometimes the behavioral changes are subtle, and pet owners often don't notice symptoms or don't report them to their vet, assuming they are normal signs of aging. Veterinarian David Taylor, author of *Old Dog, New Tricks*, mentions a few of the most common behavioral changes seen in older dogs, along with their likely physiological cause: dental disease— suffered by the majority of old dogs—can cause pain, thus irritability; loose stools can cause house soiling; less efficient lungs reduce oxygen levels, leading to decreased energy, a tendency toward confusion during the night, and senility; heart disease restricts a dog's ability to exercise and leads to more sleeping during the day (all dogs over the age of thirteen have some degree of heart disease, he says); inefficient processing of waste by the liver can contribute to cognitive dysfunction; kidney disease can cause excess urine production, which can sometimes lead to urinary accidents in the house; enlarged prostate can lead to incontinence; an underactive pituitary can lead to increased irritability, overeating, excessive drinking, restlessness, and house soiling; loss of bone density and muscle mass may decrease mobility; and failing senses can lead to increased vocalization, fear, and aggression.

One of the most common behavioral issues for older animals is anxiety, Ody's favorite neurosis (and a bugaboo for many humans, too). Taylor says of anxiety in old dogs: "Although they are experienced in life and set in their ways, old dogs often exhibit signs of anxiety that can involve problem behavior. They can become more irritated by or fearful of changes in their environment. . . ." Anxiety can often be related to physical infirmities. For example, a dog whose body is producing an excessive quantity of urine may be anxious about soiling the house. Also, the loss of sight and hearing can create feelings of anxiety, as can cognitive dysfunction. "Some elderly canine behaviors," Taylor writes, "are expressions of a conservative, averse-to-change attitude which is similar to that commonly seen in old people." Ody is most certainly anxious about his physical changes, particularly the refusal of his hind end to behave as it should. I can see worry in his face as he sways and limps and struggles to stand upright enough to eat.

Taylor believes that many of the problematic behaviors related to aging can be addressed by a committed pet owner. Perhaps the most important point is this: behavioral problems such as urinating in the house oftentimes stem from a medical condition, and the vet is the first person to see, not the behaviorist (who might be second). The first person to see is definitely *not* the euthanasia specialist or the shelter in-take worker. If we anticipate behavioral changes as our animals age, we can remain proactive in seeking their root cause and helping our animals adapt to aging. We'll also be far more likely ourselves to adapt successfully.

Of course not all behavioral changes in older dogs are adverse or hard to deal with. For instance, while many older dogs suffer from increased separation anxiety (perhaps spurred along by a heightened ambient level of anxiety), Ody's separation anxiety has actually gotten much better. Because he is mostly deaf and very often asleep, he doesn't notice when we leave and doesn't panic; because he is addled, he forgets that he is supposed to destroy things while we're away. But even when he does see us go—when he stands in the hallway and watches all the rest of us going out into the garage and piling into the car—he doesn't seem all that concerned or interested. I think he actually looks forward to us leaving because he has a chance to check all the dog dishes and then all the counters and cabinets and eat what he finds without worrying about getting in trouble with Topaz.

Behavior books talk blithely about retraining. "If your older dog is having issues," they say, "a period of retraining is called for." I sigh when I read this, knowing full well what "a period of retraining" means: another series of opportunities for Ody to be obstinate and for me to fall short. I have learned from Ody that almost all behavioral issues in dogs have a strong human component: we give confusing signals, expect our dogs to understand a foreign language, and become upset when they fail to read our minds. And I will be the first to say that training or even retraining a dog is not easy and can require a substantial commitment. Granted, it may only take fifteen minutes a day, but these fifteen minutes are somehow very hard to lock down. It is like trying to improve one's daily diet: consistency and commitment are slippery and sometimes the harder you try, in your mind, the less you accomplish in fact. But it is not only our own well-being that is on the line here.

SUCCESSFUL AGING

How well our dogs age, and even how long they live, boils down not only to their genetics but also to how we care for them throughout their lives. Consider all the ways in which lifestyle choices affect human longevity and the quality of our golden years: what we eat, how much we exercise, whether we smoke or drink, how much stress we place on our bodies and minds. Every day we make choices that have long-term consequences ("Would you like fries with that burger?") Within the human-aging literature the concept of "successful aging" became popular in the 1980s. Aging successfully means keeping your odds of developing disease or disability low (through healthy behaviors such as exercise and eating lots of veggies), keeping your cognitive functioning high (by doing brain exercises, playing chess, continuing to work), and staying actively engaged in life. Aging successfully also involves adapting to age-related changes and learning to live with disease and disability. Although I haven't seen any references to successful aging in animals, it seems appropriate to apply this idea to our companions. Our goal, as caregivers, should be to help our pets age as successfully as possible, given the constraints of individual genetics and uncontrollable environmental factors.

Another way to look at this is with the distinction gerontologists make between primary and secondary aging. Primary aging refers to genetically programmed changes, which are built into the organism and more or less hardwired. The changes associated with primary aging include fading vision, hearing impairment, and a reduced capacity to adapt to stress. Secondary aging refers to physical deterioration related to lifestyle and is largely controllable. We can offset secondary aging by living healthy lives: eating well, staying fit, avoiding excesses of alcohol and tobacco, and using medical services to manage disease effectively.

For our pets, secondary aging depends almost entirely on us. If we feed our dogs cheap, highly processed food—the canine equivalent of Twinkies and french fries —and if we feed them too much, they will (like us) get fat, and this will accelerate their aging. Between 25 percent and 40 percent of dogs are considered obese, and this number continues to rise each year, in step with the rising rates of human obesity. Although most veterinarians agree that the really cheap corn-based kibble is not the healthiest for our dogs, the vehemence with which junk kibble is decried runs from tepid ("it doesn't really matter what you feed them") to the frothing mad ("junk kibble is killing our animals"). No one seems to agree on the ideal canine diet. Raw or cooked? Homemade recipes or prepackaged kibble? Organic or chemically enhanced? Carnivorous or vegetarian or vegan? Incidentally, one of the longest-lived dogs on record—a border collie who lived to be twenty-seven—was fed a purely vegetarian diet. All the disagreement about proper diet makes it hard to know whether you are actually doing the right thing for your pet. (I go back and forth between homemade concoctions—recipes drawn from *Dr. Pitcairn's Complete Natural Health for Dogs and Cats*—when I can manage it, and high-end kibble, during times when I'm having trouble finding time enough even to cook for my human family. I don't know which is really best for my dogs. But I can tell you that homemade is their favorite.)

Exercise, fortunately, is relatively straightforward. Except that for those of us who don't live on a farm or large country estate, exercising our dogs is something we actually have to *do*. Getting our animals off the couch means getting ourselves off the couch, too. We have to walk or run with our dogs or take them to a dog park or some other area where they can stretch their legs.

Many people take a hands-off approach to their pet's health: if it ain't broke, don't fix it. The problem with this is that often things are going wrong inside long before the animal reaches the point of medical crisis, and not only is the animal now suffering, but there may have been simple and effective ways to manage or even prevent the problem in the first place. I used to be a bit skeptical of those recommended wellness visits, which were, I suspected, just a way for the vet to squeeze more money out of my pocketbook. I figured that if Ody looked fine he was fine. But I usually went anyway, albeit begrudgingly, because I felt that I should, and I'm the type of person who likes to follow rules. But of course he wasn't fine, as a full blood panel revealed.

Whether we spay or neuter can also impact a dog's health as he or she ages. Generally speaking, spaying and neutering improve a dog's chance of remaining healthy. We also, in various small ways, have to watch over the safety of our animals: making sure they are not loose near roads; seeing that they don't eat our socks, don't choke on chicken bones, and can't reach the chocolate truffles off the counter; cleaning their teeth at regular intervals; removing knots and burrs from their coats; and so on and on. It sounds very much like caring for a child, doesn't it?

In addition to the trifecta of pet health—diet, exercise, and veterinary care—we could also list various "intangibles" that can influence how our animals age, such as levels of stress, anxiety, pleasure, happiness, social contact, mental stimulation.

Those of us who try to live healthfully know that health isn't something that just happens; you have to work at it constantly, and you need a certain level of self-control and self-discipline. With our pets, too, health requires careful thought, planning, and daily discipline. Most of us know from experience that there is no such thing as perfect, but we owe it to our animals to do the best we can for them.

ELDERCARE

If we want our animals to age successfully, we have to begin long before they are actually old. And once they are old, there are a great many things we can do to help them stay functional: keep them mentally engaged and physically active and make sure they have meaningful social interactions with people and other animals. In addition, there are a great many adaptations that we can make to keep our aged and infirm animals enjoying life for as long as possible.

My experience with Ody is a lot of "could have/should haves." I could and should have done more to prevent secondary aging. But now that he is old and somewhat disabled, I am learning that I can still be proactive about maintaining his quality of life. I find myself needing to ask: In what ways is he compromised and is there anything I can do to help? I have been surprised at the variety of resources available for elderly and disabled pets. As I research, I find many ways in which I can and should be helping Ody live out his final days as comfortably as possible.

Take, for example, vision. As with aging people, companion animals often lose visual acuity as they age. And I think Ody's visual losses also make him concerned, and I am ashamed of how long it took me to realize this. I thought I was doing Ody a favor by letting him trail along behind me and Maya and Topaz when we went for our nightly walk up the street—Ody was leash free, as a special reward for being so pokey and old. Still, he often refused to go with us and after reaching the patch of evergreen bushes at the end of our yard he would turn around, stumble back to the screen door, and wait for us to return. It took me far too long to realize that Ody's reluctance might have been related to his compromised vision. We got too far ahead and he couldn't see us, so he simply returned to the safety of the back door. After pounding myself on the head for being daft, I began leashing Ody again and keeping him right by my side. And voilà: he happily walked up the street with us.

Ody was probably losing his hearing long before I recognized it, and I wonder how many times I unfairly accused him of ignoring me when I yelled "Come!" I realize now, having educated myself a little about deafness in dogs, that hearing loss is quite common in older animals, just as it is in older people. For Ody, the main impact of his impairment has been increased social isolation. And I must admit that I do not always take the extra steps that I should to encourage Ody to participate. Sometimes I let Ody stay asleep on the couch as I leash up Maya and Topaz for a walk because it is simply easier to take just the two of them. *He won't even notice*, I think to myself. *He'd rather sleep.* But this is really unfair.

A diagnostic test called the BAER (brainstem auditory evoked response) is available in some veterinary hospitals. The test detects electrical activity in the cochlea of the ear and in the auditory pathways in the brain and can give a sense of the extent of hearing loss in a dog. (Some breeders use the BAER test on pups—particularly in breeds like Dalmatians, Australian shepherds, and Great Danes in which coat-color-related deafness is relatively common. Unfortunately, the test is often used to determine whether a pup will be euthanized. Deaf = defective = dead.)

For animals who, like Ody, have decreased mobility, there are a variety of possible ways to help. For example, at handicappedpets.com, you can buy doggie wheelchairs. These are designed for dogs who have full use of their front legs but whose hind ends no longer function because of arthritis, hip dysplasia, paralysis, or some other disabling condition. A harness attaches around the chest of the dog, and two supported wheels hold up the back end. Dogs can walk, run, swim, pee, poop, and generally smell the roses. There is also a cart designed to support both front and back limbs. And you can buy support leashes and harnesses that the owner holds while the animal walks.

At this point, Ody does, in fact, need help walking. And I've debated whether to order him a cart (I found a couple used ones on ebay for reasonable cost—retail price is a whopping $500). But I ultimately conclude that it wouldn't be good for him. His front end isn't that strong, it would be very stressful just putting the cart on and taking it off, and the cart would really only help him walk, not help him get up, which is his real problem. I can envision him bumping into things all over the house (like when he got stitches and had to wear an Elizabethan collar) and becoming more and more stressed out. It would confuse his already compromised sense of where his body is in space.

Sometimes simple modifications to one's home environment can help an elderly pet move around with more confidence. For Ody, some of the most useful modifications involved placing carpets and mats over the wood floors to decrease his chances of slipping and to help him feel more secure. We also tried building a ramp so he could more easily move in and out of the dog door. And we tried various kinds of ramps and steps to help him into the car, with varying degrees of failure.

Handicappedpets.com also sells a doggie version of Depends. As Ody begins to have occasional accidents in the house, I wonder if I ought to try some dog diapers, which you can buy in a variety of fun prints and colors. The website tells us:

> Leaky pets are often kicked out of bed, locked in the laundry room and left at home when visiting. NOT ANY MORE! We have Pet Diapers for dogs and cats, Male Wrap Belly Bands, Suspenders and Fancy Pants to keep diapers on and give them pizazz, and (what will they think of next?) Bird Diapers! "Buddy is so happy that he can sit on the couch again and sleep with us in bed!"

In addition to the Fancy Pants, and to a choice between disposable and washable pet diapers, you can also purchase a "drag bag" (use your imagination). I make a mental note of this website and the various products. I am touched and also a bit horrified. I'm not sure Ody needs any more pizazz, and I have a hard time getting past my sense that diapers would be degrading, though I know this is *my* hang up, not Ody's. So far, his incontinence is infrequent. Our real problem is that once he poops in the backyard, he steps in it and tracks it all over the house. Unfortunately, diapers won't help with this.

Ody is pretty slim, despite how much food he steals. But if he were slightly overweight, I see at PetSmart that I could purchase him the Hill's Prescription Diet r/d Weight Loss System, which has a picture of a smart-looking beagle on the front, her svelte belly wrapped with a blue measuring tape to show just how perfectly proportioned she is. The large boxed kit comes with sixty-six prepackaged low-calorie meal and biscuit servings, as well as a guide that includes tips for success. If Ody had actually become obese from too many hotdogs and bags of flour, I could purchase the more serious Hill's New! Therapeutic Weight Reduction Program. Also, I could enter him in the online Hill's PetFit Challenge. If none of these things worked, I could ask the vet for some Slentrol, the "first prescription weight-loss medication for dogs," courtesy of Pfizer. Unfortunately, Slentrol appears to have some unpleasant side effects, such as vomiting, diarrhea, and lethargy.

In addition to the various doodads you can buy to help your aging pet, you can also pursue various veterinary treatments to increase functionality. Steroids can help some animals regain mobility; others respond well to glucosamine; and still others to tramadol. Joint replacements are becoming more common, both for working dogs and companions. A *New York Times* article about joint replacements for dogs competing in agility told the story of Lily, a nine-year-old pug who went from being crippled to once again competing in agility competitions, after a total hip replacement. One of the avenues that I wish I had explored more carefully for Ody are the nondrug therapeutic alternatives such as canine rehabilitation, physical therapy, massage, laser therapy, and acupuncture.

CAREGIVING DILEMMAS

I realize that not all pet owners see their companion animals as members of the family. Dogs can serve purely functional roles, for instance, as security system or hunting tool, and even within a family setting, many dogs are not treated as true companions. But in many households, companion animals are considered a true part of the family. Indeed, a recent survey reported that 81 percent of respondents consider their dogs to be equal in status to children, and more than half refer to themselves as pet parents and call themselves their dog's "mommy" or "daddy." Three-fourths of respondents admitted talking to their animals as if they were family members, and many pets have a slew of endearing nicknames. People celebrate their pets' birthdays, buy them Christmas presents, and have special scrapbooks to record all of Scrappy's extraordinary achievements. (I am guilty of every single one of these things, except for the scrapbook.)

Pets clearly occupy a special place in most families. But what exactly this place is comes under special scrutiny when an animal begins to have significant needs. What is the best analogy for the role that elderly animals play within the family? Is taking care of an elderly dog similar to taking care of a child? Or is it similar to caring for aging parents? I suspect that within each family with animal companions, the role differs, and there doesn't seem to me a clear right or wrong (though I do think, honestly, that some people go a bit too far with the dog-as-child thing). Although we joke about Ody being our first born—he came along right after Chris and I were married, and about two years before our first human child—we don't really see him as a mutant child. And although Ody's aging process has a good many parallels with the aging of my parents, I don't see him as a mini adult, either. Dogs just are what they are—their own beings—and they are fully a part of a family system. I think of a family system as a unit trying to maintain homeostasis, and ours just happens to be a multispecies unit.

An ailing pet can certainly create tensions within a multispecies family system. One way to avoid this is simply to opt out: euthanize the animal before her or she really causes problems or in some way inconveniences the family. Another option—the one I favor—is adaptation. But this, I know, is easier said than done. I have discovered deep ambiguities and ambivalences about how to balance Ody's needs against the needs of my human family, against the needs of Maya and Topaz, and against my own needs (how much of oneself must be given over to the caregiving task?). There are no ready answers to these questions and also no perfect balance to be found—always there will be uncertainty and guilt. And, as I have experienced it at least, the role of caregiver is never static: each day is different— Ody's needs, my needs, the family's needs are constantly shifting.

I thought it might help me elaborate my role as caregiver to look through some of the human-caregiving literature, particularly the literature on caring for elderly parents. *The Emotional Survival Guide for Caregivers* includes topics such as defining commitments, utilizing support, handling sacrifice, weighing hope, acceptance, fantasy and reality, fostering awareness and flexibility, protecting intimacy, and sustaining the spirit. The most important point I glean from this book is an obvious one, but one that I hadn't really considered: caregivers are problem solvers whose role is a proactive one. We are not passively comforting but actively seeking solutions. I also realize that the most perplexing set of challenges may be emotional. The survival guide says that caregivers may experience anger, sadness, guilt, and frustration as well as deep satisfaction, compassion, and love.

Guilt. No matter how much you do, it's never enough—not now, nor has it ever been. I have always been good about vaccinations and yearly wellness checkups, but have I been proactive enough? One book about caring for elderly dogs recommends taking your dog to the vet every three months for a complete workup. I didn't do this. I am not consistent with Ody's fish oil. I run out and don't get back to Vitamin Cottage for a week or two or more, or I forget. This is partly because I don't see that it helps a great deal, but it is also, I must admit, because I have other things on my mind. I feed Ody hotdogs and lunchmeat because he loves them, but I know these can hardly be called nutritious. And his teeth. The condition of his teeth is partly Ody's fault, for eating so darn many doors and couches, but I also was inconsistent with his dental care, brushing his teeth with the nasty chicken-flavored toothpaste only when I remembered (which wasn't nearly often enough), and sometimes avoiding the task simply because Ody hates it and this makes it hard. I wonder, too, if we ran him too much when he was young—like the day we rode our mountain bikes over the twenty-plus miles of the Gemini Bridges trail in Moab, with Ody alongside us—and if the deterioration in his back legs would be less severe if we had asked him to do less. Ody has lived well past the 12.5 Vizsla average, so maybe we did okay. But couldn't we, shouldn't we have done better? Could have and should have. This phrase is full of regret, of things not done or not said. And for me, it encapsulates the role of caregiver for an elderly animal.

HUMILIATION

Ody fell over in the kitchen one day, after Topaz tried to nip at his Achilles tendon. He was frightened and distressed, and as he slipped and struggled on the bamboo rug, he began to poop. As he flailed, the poop smeared on his back. When I reached down to help him up, he tried to bite my arm. Finally, I managed to lift his soiled hind end and he tottered off out the door, not once looking back. I felt, for Ody, the sting of humiliation.

As I thought about this event later, I wondered how closely my own reaction tracked onto Ody's. Did Ody himself feel humiliated? Or was I projecting my own feelings onto him? I suspect the latter, but I'm really not sure. Charles Darwin claimed that shame and embarrassment are uniquely human: no other animal has the physiological means to blush. Yet Darwin's insights only go so far, and I suspect there is much about animal emotion that we still don't understand. Perhaps animals do experience something akin to humiliation, shame, or embarrassment. Perhaps they don't.

And this leaves me feeling a bit unsettled. One of the most commonly cited cues that euthanasia may be appropriate for an elderly dog is when he can no longer urinate or defecate cleanly, when he has no choice but to go where he is, and we find him, in the morning, lying in a pool of his own urine or covered with his own feces. What you hear people say about situations such as these—and what I myself felt as Ody struggled on the floor—is that it is humiliating, and this is what is so bothersome. Because we don't know whether animals feel humiliation—and let's say, to be conservative, that they do not—then we need to look more closely at the logic behind our own concern. If we alone feel the sting of humiliation, should this so strongly influence our decision making about when to euthanize?

Merely being covered in poop would not, I can attest, bother Ody in the least. He has emerged from the bushes on many a hike covered from nose to tail in the excrement of some animal or another, and he bears a look of great satisfaction. Being covered in his own excrement may be different—I really don't know. But although I don't think Ody was feeling humiliated during the fecal incident, he was clearly distressed. What might be distressing to an animal about soiling itself is a sense that they have lost the capacity to engage in one of the most primitive of the "natural" behaviors (do not soil your own nest). Or, as with Ody, the distress may come from not being able to stand up and move around. Or it may be that they feel anxious about doing something they have been trained, from puppyhood, not to do, and they don't like to disappoint us.

I'm not sure that animals feel humiliation (as a noun), but it is certainly true that we can *humiliate* them (as a transitive verb). A particularly memorable humiliation occurred during a camping trip in Moab. One evening, our always-looking-for-a-laugh friend Max taped a huge sausage to Ody's belly, like a super-long penis. Max then sent Ody traipsing through the campsite as we all sat around the fire, causing a huge uproar. Ody seemed oblivious to the joke and was simply happy to be the center of attention. Yet there was something about the whole episode that felt wrong because the joke was at Ody's expense; it humiliated him in our eyes. (The worst part for Ody was when we had to pull off the tape, along with a substantial helping of red hair. Ody had his revenge the next day when he stole the remaining sausages off the table and Max had no dinner.) If you think about how we relate to and use animals—take a walk down the greeting card aisle of your local store or go see the circus—it seems we humiliate them all the time.

HOW MUCH IS TOO MUCH?

The field of human bioethics has been much more about saying "no" than about saying "yes." An ever-expanding array of new treatments and technologies are on offer, and once something is available, it is very hard to say no, especially with medical treatments. Isn't it more loving and more loyal to our aged father or ill child to tell the doctors to do everything and leave no stone unturned? Yet as people who become embroiled in these technologies have discovered—doctors and patients alike—doing more can seem futile, irresponsible, and even cruel.

We may be following a similar arc with our pets. The options available for treating seriously ill animals are expanding, and we now have many of the same choices that we have with people, which makes it all the more difficult to say no. I can feel this pressure already, with Ody. I could, perhaps should, have looked into his vision problems. Maybe there are corneal implants that would sharpen his sight enough that he could still catch treats in the air. Or I could pursue diagnosis and find out if he has liver cancer or bone cancer. I could do more blood tests, an ultrasound, a biopsy of the liver. The vet has presented these options to me. We could resect a portion of his liver, try to salvage what's left. We could do an X-ray and see if there is spinal impingement, and if there is, she could operate on his back. But a simple blood draw is stressful for Ody, even when it is done in the comfort of our home. With Ody, I simply say, "He's too old." Is this wrong? Is old age an appropriate reason for saying no to potentially helpful treatments?

Japan has reportedly opened the first-ever nursing home for dogs, and I wouldn't be surprised to see something like this come to the United States. People pay the equivalent of about $37,000 a month to house a beloved dog at the Soladi Care Home for Pets, which has an assisted-living program—with round-the-clock monitoring—run by veterinarians. In addition to medical care, each dog receives a specialized program of physical therapy and exercise. They are fed a special diet and are even given time to play with a group of puppies, to help keep them feeling spry. The downside, of course, is that the dog is no longer with his or her family in the place he or she knows as home. Is this going too far?

OLD ANIMALS IN SHELTERS

Rooby stares at me, her gaze following me as I move closer. When I reach the glass, her bushy golden tail begins to wag. The Longmont Humane Society is busy this Saturday. A crowd congregates around the puppy enclosure, and I hear over and over the "oooh" and "aahh" and "oh, how cute!" A trickle of people move past one enclosure and then the next. They slow as they pass most dogs, but not for Rooby. People's glances slide over her as if she were an empty cage. While I am there, no one takes her to a visitation room. No one even reads the little sign that tells her name and age and that "this sweet girl is looking for a forever home." Her gaze follows each passing person. Does she know that she hardly exists? Is it because she is twelve, and undeniably an old dog? Do her scruffy fur, graying muzzle, thick waist, and cloudy eyes betray her?

The term "ageism" was coined in 1968 by gerontologist Dr. Robert N. Butler, who noted a widespread and insidious prejudice against elderly people. (Butler also founded the first department of gerontology at a US medical school, at Mount Sinai Medical Center, in 1982.) Are we ageist toward animals, too?

It is a sad fact that millions of elderly companion animals spend their final days in a shelter. Some of these senior dogs have lost their elderly human companions to old age and disease. Rooby, for example, was brought to the shelter because her elderly owner had passed away and none of his remaining family wanted her. But far too many of these seniors have been surrendered simply because they are old. And too few will find new homes. I've heard many stories about people who go to enormous lengths to care for a senior dog or cat—sometimes taking several months off of work or even losing a job. Yet this is not an option for everyone. Sometimes people are genuinely backed into a corner, emotionally or practically or financially, by the needs of an elderly animal. And sometimes, unfortunately, a person approaches his or her animal with nonchalance or indifference: once it becomes too threadbare or faded, you just take it to the recycling center or the dump. Even people with the financial means are often unwilling to ante up for the increased veterinary bills, medicine, or special food that a senior might need.

The Grey Muzzle Organization compiled a report on old dogs and animal shelters, trying to understand how senior dogs wind up homeless and what can be done to make their lives more comfortable and facilitate their adoption into a new home. Just how many elderly dogs wind up in shelters is unknown, but every shelter in every town is likely to have some old dogs. One of the most common explanations given for surrender of an old dog on a shelter intake form is "moving." In "shelterese," "moving" is a euphemism for "I can't be bothered anymore. I'm moving on."

Sometimes it is medical needs that drive older dogs into shelters. Medical expenses tend to be higher for older animals, and some owners may be either unwilling or unable to provide adequate veterinary care. For all the seniors that end up in shelters, as many or more will simply be taken to the vet and euthanized. Having a vet euthanize an animal is more expensive than surrendering it to a shelter—about $200 compared to about $40— but it probably feels, to some, like a cleaner, more emotionally acceptable choice, less abandonment than aggressive medical treatment for old age. Sometimes, too, old animals are dropped off at a shelter because they need to be euthanized, and the euthanasia and disposal fee is too much for the owners. Burt, who ended up at the Sanctuary for Senior Dogs, was found lying in a ditch, his body riddled with buckshot.

Health problems for old dogs in shelters are a challenge. According to Fred Metzger, a veterinarian who specializes in senior animals, dental disease is epidemic in older pets. Unfortunately, it is expensive to treat. Most senior dogs would benefit from regular blood work because it helps cue vets to the presence of various common diseases such as kidney disease, hyperthyroidism, and diabetes. But, again, blood work is expensive and beyond the budget of most shelters. Untreated health problems make adoption even more unlikely for senior dogs. It isn't until these animals are in a medical crisis that they see a shelter vet, and by then the best and most financially viable option is usually euthanasia.

Luckily, among all this elderly heartbreak, there are people who have a special fondness for elderly animals. A small but growing number of community animal shelters are actively seeking to increase the proportion of seniors adopted. For example, the Elderly Animal Rehoming Scheme, by the Royal Society for the Protection against Cruelty to Animals in the United Kingdom, seeks to improve the odds for elderly animals to find new homes by helping people overcome some of their fears about adopting an elderly pet. The RSPCA educates potential owners about the realities of older pets, they promise discount vet visits, pledge help with transport, and offer a twenty-four-hour emergency telephone line. Shelters will often hold a "September Is Senior Month" event, aimed at not only raising the profile of the elder dogs at the shelter but also encouraging adoptions through special pricing and incentives. The Longmont Humane Society, in Longmont, Colorado, gives regular discounts on all senior adoptions and holds a special senior adoption event at least once a year.

A number of rescue organizations make it their specific mission to help older animals: Muttville, Old Dog Haven, the Senior Dogs Project, Sanctuary for Senior Dogs, BrightHaven, Furry Friends Haven, A Chance for Bliss. Some of these, such as Old Dog Haven—whose newsletter proclaims "We Love Old Dogs!"—are networks of private individuals helping homeless senior dogs by providing foster homes and actively seeking adoptive families. Others, such as BrightHaven, serve as sanctuaries where old or disabled animals can go live out their remaining time. There are similar programs for older cats, such as Purrfect Pals and Tabby's Place.

Some shelters have special programs (called "Seniors to Seniors" and the like), which aim to find mutually enriching partnerships between elderly animals and elderly humans. Although the relationship between pet ownership and the well-being of the elderly is complex and the data inconsistent, some studies have shown that elderly people with pets remain more active and have lower levels of depression than those without. Elderly animals can enrich the lives of elderly people in other ways, too: Many retirement homes and hospice centers incorporate pet therapy programs, and some, like the Steere House where Oscar the death-predicting cat lives, actually keep pets on the premises because having animals adds (ironically) some humanity to a place, making it more homey. Animals can sometimes connect with people on a level that other humans cannot, and even people who have never really known or even touched an animal can be very moved by visits from a dog or cat. Elderly animals often make the best therapy pets because they are calm and self-possessed.

More and more people recognize that elderly animals need care and protection, and this extends beyond cats and dogs. Aging horses can live out their lives at the Retirement Home for Horses, in Alachua, Florida. Aging elephants are provided for at the Elephant Sanctuary in Howenwald, Tennessee. And chimpanzees retired from medical research live at Chimp Haven, in Keithville, Louisiana. Just as retirement homes for people can run the gamut from fabulous to frightening, these homes for elderly animals are not always safe havens. The *New York Times* recently ran a piece about the Thoroughbred Retirement Foundation, one of the largest private organizations dedicated to caring for retired racehorses. The foundation failed to make payments for the care of over a thousand horses, and many of the animals under its care wound up starved, neglected, and, in some cases, dead.

I imagine old Ody sitting in a cold concrete kennel in a shelter, with a cacophony of barking and banging and yelping in the background, the smell of feces and fear palpable in the stale air. His anxiety monitor would be on high alert, Code Red. Would he look appealing to a prospective adopter? Doubtful. He would pant in her face and she would get a good whiff of old man breath. If she stroked his back, her hand would gather a little clump of white flakes and shedding hair, she would feel all the lumps on his belly and see the skin tags of various colors hanging off elbows, cheeks, haunches. He might try to bite her, too. Someone would need to see past all these blemishes. They would need to look into his cloudy eyes and see the real Ody, the dog of intense appetite, boundless personality, and a capacity for love as big as the Wyoming sky.

But I don't like this thought experiment. It is too heartbreaking. In my ideal world, each and every senior dog would have a warm and loving home in which to live out their last days. Each and every Ody would be recognized for who he really is—not just an old dog. And if, after Maya and Topaz have passed away, I ever adopt another dog, I promise I will go to the shelter and find the oldest, smelliest, scruffiest dog I can.

The Ody Journal

JUNE 5, 2010

Even though Ody stumbles and limps through his walks, and has a hard time holding his rear end off the ground, he still, incredibly, manages to jump onto the various couches in the house. He never sleeps on the beds anymore—less as a matter of physical limitation than territorial concern. The beds are clearly Topaz's. But couches—this is where he spends most of his time. Ody is so quiet that I have to search the house for him (unlike Topaz who is always there, an appendage, really).

JUNE 26, 2010

As I was walking up to the park with Maya and Topaz, Bill, the neighbor who lives along the alley, called out, "How's Ody doing? You think he'll come see us again sometime?"

"Ody's hanging in there," I reply, "doing his old man stuff. He doesn't get out much anymore, though."

When we first moved to this house, we had a heck of a time keeping Ody in the yard. One of our first home projects was a fence. This was high on the list because our house is on a somewhat busy road, and with a dog and a small child we felt nervous having an open yard. We built a nice six-foot wooden fence around three sides of the yard, but along the back, where our property meets the neighbor's, a row of beautiful Spiraea bushes grow along the border. In the spring, the Spiraea burst forth in an explosion of white blossoms, and year round, the greenery creates a nice feeling of privacy. To keep the Spiraea, we had to install a four-foot chain-link fence.

Right at the corner, where the chain-link meets the wood, was Ody's favorite place to go. We could stand at the sliding glass door and watch Ody make his escape. First, he would look around to make sure we weren't watching, turning his head back over one shoulder and then the other, very sly-like. Then he'd make his move. He wouldn't jump over the fence like a normal dog, nor would he scrabble over the way a smaller dog might. The only way to describe his method is as a slither. He would hook his front paws over the top, and then shimmy his way up and over.

Was Ody unhappy with us? Is this why he tried to escape? I don't think so. I think his motivation was just to seek out more human companionship—to find the love of strangers. No matter how much love we showered on him, maybe it wasn't enough. He had some hole that we couldn't fill.

Clued in to his penchant for fence jumping, most of the time I caught Ody in the act. Sometimes I found him half slithered up the fence, and at my sharp reprimand he would let his body ooze back to the ground and give me one of his best guilty looks. But a few times he did truly escape. On one of these adventures, Ody went to visit Bill. I got the call one spring evening, having just arrived home from work. "Do you have a red dog, goes by the name of Odysseus?" he asked me. "Yeah," I said. "That's my boy." "Well, he's been up here hanging out in my garage. He came wandering in to say hi, and I didn't want him to get hit by a car, so I just held on to him." When I walked up to retrieve Ody, he didn't want to leave. And I think this made Bill feel just a little bit attached to Ody.

JULY 3, 2010. ESTES PARK.

Arrived at the cabin today for a week-long stay. It is late and I am tired after a long day of packing, arriving, unpacking. Having finally settled everyone in for the night, I snuggle down into the bed, relieved finally to be here in the peace and quiet. Before I've even turned out the light, Ody begins to bark. After a few minutes of this, I walk out into the living room, a small cloud of annoyance hovering over me. Ody is just lying in the middle of the floor, looking at me. He isn't waiting by the door to go out, and anyway, why would he need to go out? We had just gotten back from our evening walk. So peeing can't be the issue. Is he hungry?

I do what I know I shouldn't. I feed him a second dinner. I know I shouldn't, because he may then associate barking with some reward. And I certainly don't want to encourage the bark. But at the cabin, the dogs really do work up an appetite, and anyway, I'm really anxious just to climb into bed. So Ody has a second helping (and so do the other dogs because they are always on guard to make sure everything is fair).

Back into bed, I turn out the lights and settle in, again. Quiet. Then it comes. The first bark. Silence for one or two minutes. Second bark. Third. Sympathetic Me feels sorry for Ody—he's restless, lonely, disoriented. My heart breaks a little bit for him. Unsympathetic Me feels sick and tired of the incessant barking, of never, ever sleeping uninterrupted for more than a couple of hours, of having to drag myself out of bed every night at 4 a.m. Have I ever wished that Ody's end could come a little sooner? This thought crouches at the back of my mind, a vampire idea, unwilling to show itself in the light.

I ignore Ody for about fifteen minutes, hoping either that he'll settle down or that someone else will deal with the problem. Finally, I throw off the warm covers once more.

Feeling cross and ready to scold, I walk into the living room, expecting to see Ody in the same spot. But he's not there, nor is he on the couch or in the kitchen. Finally I find him in a heap on the bathroom floor, struggling to stand but unable to find any traction on the slick Pergo. Maybe he came in the bathroom to lie on the cool floor, or maybe he fell after taking some refreshment from the toilet.

I lift his rear end gently and lead him out of the bathroom. I pick him up now and put him on the couch, his favorite place to sleep. Finally, I crawl back under the covers, heavy hearted. All is quiet.

JULY 4, 2010. ESTES PARK.

On the hike today (which was about two hundred yards long), Ody is noticeably slower than he was even the last visit to the cabin. He fell over once, after stepping wrong on a rock. If no one is paying attention, he'll disappear on little adventures. Just trudging along in the wrong direction on his inscrutable business.

JULY 7, 2010. ESTES PARK.

Took Ody and the others on a longish walk last night—all the way down to Hell's Hip Pocket Ranch. This is really only about a third of a mile, but for Ody, quite a long way. His left back leg seems much worse. He hardly picks up his foot but, instead, sort of drags it along. And the whole leg seems turned out at a funny angle. His body twists to the side, and every step or two, his back end collapses. This used to take place every ten or fifteen steps; now it is every other. But he still seemed to have a nice time. Quality of life intact.

He actually slept on the bed with me two nights ago. At home, he never sleeps on the bed. But here at the cabin must feel different; it must not feel quite so much like Topaz's territory.

When I pick him up to carry him up or down the stairs, or to put him in the back of the car, he often tries to bite me. He reaches around with his head but can't really get it very far, so he's just curls his catfishes and bares his stubs of teeth. The other day he actually managed to bite my arm, but he has little power in his jaws and no teeth to speak of, so it didn't really hurt.

According to Sage:

Ody's totem animal is the water buffalo.
Maya's totem is the deer.
Topaz's is the Tasmanian devil.

JULY 8, 2010. ESTES PARK.

Ody has a new growth, a small red bulbous thing on the end of his tongue.

Had a nice love-fest with Ody tonight. I was comfortably curled up in the armchair reading by the fire, and there he was lying in the middle of the floor. It was one of those "live in the moment" moments; I thought to myself that Ody needs to know he is loved—he is so often left behind, left out, left on the sidelines. And how many more chances will I have to show him how much I love him? I curled my body around his, and stroked his fur from head to tail, feeling down each of his front legs. And I just hugged him for a long time. Finally Topaz, overcome with jealousy, stuck his nose between Ody's face and mine and broke the spell.

JULY 8, 2010. ESTES PARK.

I took Ody on a walk down the road. It was a hot day, and I wanted so much for Ody to be able to go down into the creek to cool off. Lolling around in water has been one of Ody's great pleasures in life. He loves to stand in chest-deep water, but he most definitely does not like swimming. He didn't learn to swim properly until he was about ten, and even then he paddled his legs so hard he created a turbine in the water. To coax him to swim at Boulder Reservoir, Sage and I would tie a string around a hotdog and pull the meat through the water. Ody would inch his way out toward our aquatic lure, his desire for hotdog slowly but surely overpowering his fear of deep water.

I finally did find a place to take Ody down to the creek, though I had to help him down the slope. He seemed to enjoy putting his feet in the cool water; he took a few licks. But then his feet got tangled in a submerged branch and he ended up falling backward into a sit, from which he couldn't get up. The whites of his eyes flashed at me.

Later, when I was off walking Topaz and Maya, Ody managed to let himself out of the cabin without Sage or Chris noticing. Ody must have pushed the screen open. It makes me cringe to think of him attempting those hard cement cabin stairs by himself—I've watched him fall three times now. Maybe he went down the long way, traversing along the hillside over to the neighboring cabin, and then back toward ours. He evidently had some kind of mishap on his adventure. He's got a little bloody patch where the fur is missing on his head.

Several times while up at the cabin I walked through the Wigwam meadow, where Chris and I want to have our ashes spread. Ody can no longer make his way up the slope from the road and into the meadow. He stays on the road, carefully navigating the deep ruts.

JULY 11, 2010. HOME.

Ody seems to have aged a hundred years today. His back legs are noticeably worse. He still somehow got up on the office couch but had a hard time lifting his body to get down. His back feet kept crossing over each other as he tried to walk, and the rolling over on his toes seems worse. His whole body seems to be torqued to one side. When I look straight into his face, his eyes are cloudy white discs. What worries me the most, perhaps, is that he has begun to refuse food. He eats only bits and pieces of his dinner and no longer lurks around the kitchen when food is cooking. My canine food vacuum is broken.

Even Liz, who saw Ody two days ago and then again today remarks on the difference. "Oh my God," she says when she walks in the back door. "Look at him . . ."And I can see tears in her eyes. Big-hearted Liz has known Ody for ten years and loves him dearly. Liz's emotions lurk just beneath the surface and are easily called out. When I see her eyes, I tear up, too.

"How will you know when it's time?" she asks me. And then she says something about this week, but I don't quite hear her. Ody has lots of time.

I'm thinking in terms of months, maybe a year even. Not weeks, certainly not days. But the way Ody looks tonight scares me and I go to bed feeling a bit sick.

JULY 14, 2010

Ody has rallied. He seems perky again and ate his whole dinner (and Topaz's dinner, too).

Why obsess about physical beauty? A flower reaches its pinnacle of beauty when in full bloom, but it is after the bloom, when the flower finally goes to seed, that its potential is fully realized. This is when it is ready to pass on its secrets to the next generation. Ody has lost much of his physical beauty. As a matter of fact, someone taking a close look might just say "ewww" to Ody's various black globules, skin tags, and red sacs. But he is still realizing his potential.

JULY 30, 2010. ESTES PARK.

We went to the cabin for the weekend, partly to escape the blazing heat in Longmont (96 degrees). Had to squeeze Topaz and Ody in the back with some stuff for the remodel, and Maya had to ride up front on Sage's lap. About halfway, just after Pinewood Springs, Sage and I started to notice a horrible smell. It was definitely related to dog fart but much worse.

I didn't want to stop to check things out until we got to the cabin because it was so hot I just wanted to get the dogs safely there, and I didn't want to open the hatch on the busy mountain road and risk Topaz or Ody jumping out. So we just drove with the windows open, trying not to breath in too deeply.

We finally arrived and I opened the back of the Pilot. This is what I had feared. For the first time ever, Ody lost control of his bowels. The blanket where he had lain is covered in urine and feces. I store the blanket by the driveway—to deal with back at home. After he's had a chance to sniff around and mark some territory, I carry Ody up the stairs and into the bathtub. The only shampoo available is an ancient bottle of Pantene 2-in-1. Ody spends the weekend smelling like cheap perfume.

Our quiet night at the cabin, where I'm hoping to catch up on some badly needed sleep:

Collapsed into bed. Ody barked for an hour. Then stopped.

3:30. Ody barked. Chris let him out to pee.

3:50. Ody began again. I let him out, he peed. Then I lifted him onto the couch and tried to make him comfy with a blanket. He jumped off. I took him into the other bedroom and put him up on the bed and tried to lie next to him, thinking maybe he barked because he was lonely. He tried to bite me and jumped down. I gave up and went back to bed, but couldn't sleep. Feeling disturbed about Ody. Finally fell asleep again, maybe about 6.

8:15. Ody barking. When I get up, two cute little chunks of poop are sitting on the couch. Oddly, they are totally odorless.

I give Ody some vanilla ice cream—his favorite sweet. He sniffs the melting yellowish clump and looks warily to both sides. It isn't until I put both Maya and Topaz out on the porch that Ody begins to lick. He doesn't gobble, which is unusual. Is he savoring the experience of cool sweetness on his tongue? As he licks, his back legs begin to sag. After a few licks, his butt is nearly on the ground and he takes a break to push himself back up. Immediately the sagging begins again.

JULY 31, 2010. HOME.

Home now. We've just fallen into bed exhausted after a long day of work at the cabin and then work at home. And, just as I've gotten comfortable, "Arouf." The Darth Vader smoke detector. And then, about ten seconds later, "Arouf." And then "Arouf." I want to cry. I am so tired and I have no idea why Ody is barking, again. He's just been outside to pee, he's had plenty of dinner, it isn't the middle of the night yet. Why, why, why?

Chris and I turn to each other in bed, and we both roll our eyes. Chris says to me, "What's his deal?"

"Who knows?" I sigh, as I shove a pillow over my head, hoping the barking will just go away. And, miraculously, it does. It is now quiet.

A few moments later, Sage flings open our bedroom door and announces, "Ody just pooped on the floor." Sure enough, when we walk into the hallway, the smell hits us like a wall of poison gas. I step gingerly around two dark clumps on the floor and head toward the kitchen for paper towels. And then I feel it, squishy under my little toe. The third clump, carefully deposited right in the middle of the doorway.

As we clean, Ody peers at us from under the piano. I crawl under and give him a hug and tell him, "Don't' worry. We're not mad at you; you couldn't help it."

I feel guilty. We heard Ody bark two or three times as we were getting into bed, but like the parents of the boy who cried wolf. . . .

AUGUST 8, 2010

Ody stops at the doorway, the threshold between safe and . . . what? Unsafe? He sniffs, as if carried on tiny wind currents is his answer to whether or not to proceed. He points his nose upward, dilates the side creases, swivels his head slightly to the left. Then he turns around slowly and walks back inside.

AUGUST 17, 2010

Ody had trouble pooping last night . . . he tried and tried, slumping around the yard in the dog-pooping crouch, leaving nothing but little dribbles.

AUGUST 23, 2010

More poop on the living room floor. Didn't even hear Ody wanting to go out.

AUG. 24, 2010

Ody stands in the doorway to the kitchen, all the time. When I pass by him, I have to turn my body and squeeze along the wall. His mouth snaps open every time I walk past, as if he sees me, in his imagination, holding out a piece of hotdog.

SEPTEMBER 3, 2010

I get down on the floor under the piano to give Ody some love. Topaz watches from across the room and begins to growl and whine and then runs over and sticks his face between Ody's and mine. Poor Ody is so nervous that he doesn't enjoy the attention.

I've gotten conflicting advice from behaviorists. One said not to intervene in pack status disputes; this only makes things worse for the subordinate animal (who is, without a doubt, Ody). He went so far as to tell me to praise Topaz after he trounces Ody. Another told me to step in and protect Ody and show Topaz who's really boss (me—ha!). What to do?

I was thinking today about Ody's love life. Ody would have to be the canine version of the Marlboro Man, or maybe Robert Redford: Incredibly handsome. Red and regal, with perfect, angular lines. A rugged, strong kind of look, with that pronounced sagittal crest on his head. His nose and eyes color-coordinated to match his coat. His rippling muscles. But his love life has been oddly thin. This, of course, is because we surgically denied Ody the possibility of real, reproductive passion. But even so, I can think of only two female dogs for whom Ody has shown more than a fleeting interest. And neither of them returned Ody's love.

Samantha was an enormous, sagging, black and tan bloodhound who walked with an awkward lopsided gait, her bum curved under at a funny angle. She had a marvelous bloodhound wail. This was in the pre-Topaz era, when I actually went to the dog park regularly. Ody was generally indifferent to other dogs. He went to the dog park for the people, sidling up to one after another until he had received some pats and strokes from everyone present. But when Samantha was there, Ody only had eyes for her. He followed her everywhere, sniffing her butt and trying to mount on her back, she trying all the while to get away. She was nearly twice his size, so Ody had to stand very straight and high on his hind legs in order to grasp her back with his front paws. She was clearly annoyed by him but was mild-mannered and surprisingly tolerant of his sexual advances. She mostly ignored him, although she would occasionally turn around with a little growl. I would stand there, embarrassed, and for the sake of Samantha's Person would halfheartedly call out, "Ody, don't be so rude." "Ody, get off now." "Ody, that's enough."

Ody's other girlfriend, from the same dog park, was a Chihuahua mix whose name had some relation to a flower—Daisy, maybe? Without question she was one of the least aesthetically pleasing dogs I've ever seen. She was about the size of a loaf of bread, and watching Ody try to hump her was comical to the point of being a bit sad. He couldn't even begin to wrap his front legs around her, so he would walk behind her humping the air, a look of perplexed pleasure on his face. I know that humping isn't necessarily a sign of sexual attraction, particularly when the perpetrator has, like Ody, had his testicles removed. But I really do think that Ody had ardent feelings for Daisy.

The only other thing Ody ever tried to hump was the head of our friend Chad, one Chris's classmates in medical school. We were in Chris's apartment and Chad was down on all fours on the floor, playing with his pet ferret. Next thing we knew, Ody had run over and grabbed Chad's head between his legs and begun to gyrate. A moment of deep, surprised silence fell over the room before Chad let out a scream and Chris and I burst out laughing.

SEPTEMBER 4, 2010. ESTES PARK.

Cabin. Ody seems older yet, but he still enjoyed a walk down the road. His hind legs are even more rickety. Unlike our neighborhood walk, which is very slow, here Ody actually tries to go fast. He even breaks into a stiff run now and then, for a few steps at a time. I hold my breath because it looks like he might topple over at any moment into the chiming bells that grow thick along the creek side of the road.

4

Pain

Pain: c.1300, "punishment," especially for a crime; also "condition one feels when hurt, opposite of pleasure," (in L.L. also "torment, hardship, suffering") (see *penal*).

I woke at 4 this morning to a dog fight. As best I can guess, since I missed the lead-up, Topaz was on the bed and Maya wanted to get up, and Topaz said no and Maya said yes, and Topaz said no way and Maya said prove it. And then there was a tornado of teeth and fur and legs and tails, the eye of which was located more or less on top of my head. In my stupor, I made the classic mistake—I tried to break it up, tried to get a purchase on one or the other to pull them apart and move them off of me. I felt a deep pain in my hand and, then, a warm wetness.

After a few seconds, the storm passed and we turned on the lights to assess the damage. The floors and wall were speckled with blood. Maya had a hole ripped open in one ear, a bleeding flap of skin hanging inside out. Topaz was fine physically but was an emotional mess. His ears were folded down, his rump hunched over in dismay, tail between his legs. He watched as I held Maya's ear in a washcloth, trying to staunch the bleeding, and my husband wiped the worst of the mess off the floor. We held Maya steady and pushed the flap of skin back into place. We'd wait a couple hours and see if a visit to the vet was in order.

Then we trundled off to Chris's office to clean out my wounds, and I am thankful, as I have been many times in the past, to have a doctor in the house. I had two deep puncture wounds on my right palm and one on the back of my hand. Chris had to flush the wounds thoroughly by squirting saline deep into the holes, which hurt so much that I broke into a clammy sweat and got lightheaded. "It's just vasovagal syncope," he told me. "Lie down on the floor." *Can you die from that*, I wondered? Once the wounds were cleaned and bandaged, he told me to turn over and I got a tetanus shot in my rear end.

About 5:30 we crawled back into bed. My hand was throbbing and I couldn't sleep. So I got to thinking about the book and about the chapter on pain. Does an animal feel pain the way a human being does? Was Maya feeling anything like what I was feeling? Do dog bites hurt dogs as much as they hurt people?

In *Dog Watching*, Desmond Morris quotes from an eighteenth-century volume entitled *The Treatment of Canine Madness*: "The hair of the dog that gave the wound is advised as an application to the part injured." When I finally crawl out of bed, I grab some scissors and clip a small section of Topaz's hair. But I can't bring myself in reality to stick it into the puncture wound, so I do a little dance and sprinkle it over my head, hoping the healing powers will do their thing.

PAIN AND ANIMAL DEATH: HUMANE END POINTS

Thinking back to my tenure on the Institutional Animal Care and Use Committee at the University of Nebraska School of Medicine, what struck me then, and remains with me now, is this: the *ethical* discussion, when we engaged in such a rarity, boiled down to animal pain. We didn't discuss whether animals should be used in research (this being beside the point for an animal care and use committee), nor did we discuss the rather more tractable question of whether pain was the only moral insult we inflict on research subjects. The only experiments that elicited some dialog, and not simply a unanimous vote (minus the subject's, of course), were those that fell into the US Department of Agriculture's guidelines for Pain Category E: "Animals subjected to potentially painful or stressful procedures that are not relieved with appropriate and adequate anesthetics, analgesics and/or tranquilizer drugs."

Most of the time, for most of the protocols approved by the committee, pain was not really a problem, and the research was given the thumbs up. Among those protocols deemed painless were those that really did involve no pain, and those that involved pain but in which analgesics were administered. In other words, if I give you a strong analgesic and then slice you open, I have not actually inflicted any pain. (A koan: is pain that is not experienced still pain?)

Death, itself, was presumably not an injury to the animal. Indeed, "[American Veterinary Medical Association]–approved euthanasia procedures not involving prior surgical procedures" are included among Pain Category C experiments: "procedures that cause no pain or distress, or only momentary or mild/slight pain or distress, and do not require the use of pain-relieving drugs." Furthermore, all potentially painful experimental procedures have what is called a "humane end point," defined as that point at which pain or suffering reaches a level considered, by researchers, to be morally intolerable and at which we are required to do the kind and humane thing: kill the animal and thus ease the pain.

Within the companion animal literature, too, pain serves as the ethical fulcrum for decision making, particularly with end-of-life decisions. The question of whether and when an animal should die—when we should actively end its life—usually boils down to pain: is the animal in significant pain? If so, then death is the humane choice, the ethical end point. Death is the ultimate pain management tool. Indeed, even the broader public and philosophical discussion about animal welfare revolves around pain. The central moral question is can they *suffer* (wherein suffering is generally equated with physical pain)?

DO ANIMALS FEEL PAIN?

What an odd question, no? The answer seems patently obvious. But to many scientists and philosophers, it is not.

Pain is neither easy to define nor easy to assess. Pain is a biological response to a stimulus and results from actual or potential tissue damage. It serves a crucial evolutionary function because, without pain, an organism would not know to avoid injury. Now, we can go no further without noting our first significant point of controversy. Have we defined human pain? Or animal pain? Or both?

Pain is a physiological event (or series of events) followed by an emotional response. The physiological event is called nociception (the perception of injury or *noci*). During nociception the body receives a signal that tissue injury has occurred or might occur and sends this signal to the spinal cord. According to the National Research Council's Committee on

Pain and Distress in Laboratory Animals, "*Nociception* is the peripheral and central nervous system processing of information about the internal or external environment related to tissue damage, e.g., quality, intensity, location, and duration of stimuli. Much of nociception takes place at spinal and other subcortical levels, as evidenced by the existence of spinal reflexes that do not produce awareness of pain." Nociception is a primitive sensory capacity and is distributed broadly among vertebrate and invertebrate species. This isn't where the scientific and philosophical disturbances arise.

The conscious perception of pain is where the picture grows blurry. *Pain* is the unpleasant emotional experience associated with nociception and requires a certain level of neurological complexity. As the International Association for the Study of Pain says, "Activity induced in the nociceptive pathways by a noxious stimulus is not pain, which is always a psychological state, even though we may appreciate that pain most often has a proximate physical cause." (Got that?) If you were a very simple organism with a tiny little brain, you could have nociceptors but still not experience pain. The National Research Council tells us that "the encoding process can ultimately result in pain, but is qualitatively different from the *perception* of pain (nociperception), i.e., in the interpretation of sensory information as unpleasant. Perception of pain depends on activation of a discrete set of receptors (nociceptors) by noxious stimuli (e.g., thermal, chemical, or mechanical) and the processing in the spinal cord, brainstem, the thalamus, and ultimately the cerebral cortex."

There was a time, in science's not too distant past, when pain was considered a uniquely human capacity. Nonhuman animals, it was claimed, simply did not experience pain because they lacked the neurological complexity. This scientific creed almost certainly went against the common-sense observations of scientists, especially those in contact with animals outside of the scientific setting (e.g., those with pets). But it nevertheless remained an important ideological scaffolding for animal research—and still, by some accounts, remains in place today, despite ample evidence of this supposition's weaknesses.

FEELING FISHY: WHICH ANIMALS EXPERIENCE PAIN?

We went hiking last weekend in Rocky Mountain National Park, and a young boy, about my daughter's age, was standing at the edge of Dream

Lake as we passed by on the trail. "I caught a fish," he crowed, giving my daughter an extralong look. "Wanna come see?" We headed over to the shore, and sure enough, he had caught an endangered brook trout (illegal to catch, yes), which flopped on the shore, its pink and brown scales reflecting the montane light. As the boy bragged about his catch, the fish struggled for air, hook in mouth. "Wait a minute," he said, "and I'll show you how to take out the hook." He grabbed the end of the hook and began to yank it out.

My husband offered, with some haste, to show the boy how fly fishermen hold their catch in the water while removing the hook. "The point of catch and release," he said kindly, "is that we keep the fish populations healthy by keeping them alive." Then he asked the boy to fetch his wire cutters, so the barb could be removed before pulling the hook from the fish's mouth. The boy said he had no wire cutters. So my husband explained that it would be better for the fish to leave the hook intact. As the boy ran to look for a knife to cut the line, I retreated up the trail. I couldn't watch anymore.

When my husband caught up I asked what had happened. "The fish is gone," he said. "Oh no," I moaned, thinking of its limp body in the grasses. "No," he said, "gone. Swam away. It'll be okay, I think." My husband had helped the boy cut the fish loose and explained why he might think about using unbarbed hooks on future fishing expeditions. This particular brook trout will simply have to adapt to life with a hook in its mouth, like an extralarge lip ring.

This little fish episode took place while I was writing about animal pain, and, incidentally, while I was in the process of reading a very interesting book called *Do Fish Feel Pain?* Fish, it turns out, are more intelligent and sentient than I had thought and even feel pain. I realized that I had long been prejudiced against fish. I like fish and don't eat them, but I had always assumed that they were rather primitive. The fish book made me realize that these creatures have far more going on inside than I realized. Goldfish, apparently, are quite intelligent and easily bored. And many species of fish actually perceive and feel pain, probably much as I do. I feel uncertain now of my assumptions about which forms of life feel what sorts of things and what kinds of things should have value in my own personal moral system.

The more I read about fish cognition, in fact, the more uncomfortable I become with the tanks in my daughter's bedroom. Though I try to create a household environment in which all life is sacred—we don't kill bugs intentionally, except for ticks and mosquitoes, which we kill out of self-defense if they are caught red-handed trying to suck our blood—some life is better cared for than other life. And our fish, I have to say, rank pretty low. We have two ten-gallon tanks with guppies, one for males and one for females (they had to be separated after we discovered that they are fruitful and multiply, a lot). Our one remaining female swims around alone. Most days she hides out behind the small black pump. I'm not sure whether fish get lonely, but I worry now about this little guppy.

The answers to questions about which animals feel pain are constantly evolving as scientists learn more about animals' physiology and their cognitive capacities. Nociceptive nerves have been identified across a broad range of invertebrates and vertebrates. Humans and sea slugs both detect physical damage and respond by withdrawal from the stimulus. But which animals have complex enough brains to process nociceptive information, to feel pain?

Strong inferences can be drawn based on physiological and behavioral reactions to stimuli. Research over the past two decades has provided ample evidence that mammals experience pain and that the capacity to feel pain extends to birds and even to fish. Fish show many of the same behavioral and physiological responses to pain as other vertebrates, such as guarding behavior and increased respiration. When fish are given morphine, these responses abate.

The data on invertebrates are inconclusive. The weight of opinion seems to be that invertebrates such as worms and flies (and ticks and mosquitoes) likely do not feel pain because they have small nervous systems and limited cognitive capacity. Cephalopods such as octopuses are more difficult to judge because they have large brains. The jury is still out on lobsters and other crustaceans. The presence of opioid peptides and opioid receptors in crustaceans has been interpreted by some as an indication of an ability to feel pain, but most scientists remain skeptical. We have many surprises in store for us, I think, as we learn more about animals and pain.

SCALES OF SUFFERING

The Palm Springs animal shelter is sponsoring an event this spring called "Claws for Paws." The annual fund-raiser will generate badly needed income for the Coachella Valley's only no-kill animal shelter.

This delectable lobster meal will be available for lunch or dinner for only $20.00 including tax for a whole 1–1/4 lb. grilled Maine lobster, homemade coleslaw, confetti rice, fresh baked dinner roll and all the condiments. Beer and wine will be available for purchase. Step it up for $25.00 including tax for steak (a 6 oz. grilled USDA Choice) and a whole 1–1/4 lb. lobster. If you can't get enough, try double lobster (2 whole 1–1/4 lb. lobsters) for only $35.00 including tax.

This is a very nice idea, to raise money for the no-kill shelter. But doesn't it also seem a bit off, morally speaking? You can almost hear those sarcastic lobsters, as they dangle above a pot of boiling water, "Hey! Thanks for caring so much about animals!"

Do pain and suffering matter more to some animals than others? Do they matter to dogs and cats but not lobsters and cows? Within a creature's own reckoning, pain, suffering, and death matter completely. But in terms of us making moral judgments about which animals will suffer, if suffer some must, then we do end up passing judgments that some animals suffer in more important ways than others. If we were forced to do painful experiments on animals, it would be better—if pain and suffering were all that mattered—to use ants than it would chimpanzees. Or, perhaps more accurately, it would be less bad.

My gut instinct is that more cognitively and emotionally complex animals such as chimpanzees and dogs have richer experiences of pain than do less complex animals like rodents and fish. But biologist Donald Broom offers a different perspective. He hypothesizes that animals with more complex brains may deal more effectively with pain than animals with less complex brains. With a more complex brain, you have more varied responses, more flexibility in behavior. Maybe fish cannot deal with pain as effectively as more complex animals and may suffer more, not less, as a result.

NONPAIN CAUSES OF SUFFERING

If you took a human being and very gently removed her from her family and from all human contact and placed her in a comfortable concrete cell—being careful not to scratch or bruise her during transport—would she be happy? Let's say she has no idea why she was taken, where she is, or what has happened to her children and husband. She has nothing to do—no books, no paper, no television, just windowless walls to stare at, and a little bit of room to move around and stretch her arms and legs. The lights go on and off at regular intervals, and she has decent but bland food which is delivered on a mechanized tray through a hole in one wall. She might have to stay in this strange place for the rest of her life. Who knows? She has no physical pain; her body is totally intact. Does she not suffer?

In the animal welfare literature, pain is often coupled with the term distress. For example, the B–E scale used by Institutional Animal Care and Use Committees to rate the relative awfulness for animals of a proposed research protocol is actually called "the pain and distress scale." It is nice that the government recognizes that animal suffering goes well beyond the pain of a surgical incision, but the definition of distress—"a point at which adaptation to a stressor is not sufficient to maintain equilibrium and maladaptive behaviors appear"—leaves everything rather vague. The list of potential stressors is endless and would include, for starters, physiological stresses such as illness, discomfort, and injury; psychological stresses such as excessive handling, social isolation, and an unpredictable environment; and emotional states such as boredom, anxiety, bereavement, and fear. Distress, in other words, is that point at which stress becomes severe enough to cause suffering. (Distress comes from the combination of the Latin prefix *dis* "apart" with *stringere*, "draw tight, press together"; it is a tearing apart of that which should remain pressed together. The psyche?)

"Suffering" is an unscientific term, but it is an important part of our pain vocabulary. To me, the notion of animal suffering has more traction than the more ethically ambiguous "animal distress." Animal and human suffering are not equivalent, but we can infer from our own experiences that many varieties of suffering are available to animals, beyond the physical pain and distress caused by injury or illness. When we are making quality-of-life judgments about animals—asking whether continued life is too painful—we must surely consider a broad range of suffering.

This may seem quite obvious, but remember that skepticism about animal consciousness—and animal suffering—is alive and well. Careful scientists and philosophers will remind us that we cannot know for certain that animals suffer or what forms this suffering might take. Indeed, even the definition of "suffering" is slippery. Animal behaviorist Marian Stamp Dawkins defines suffering as "experiencing one of a wide range of extremely unpleasant subjective (mental) states," and this, she says, "is about as precise a definition as we are going to get."

Dawkins says that there is an explanatory gap between observable behavior and subjective experience—a gap that exists for humans and animals alike. "To bridge that gap, we each have to make what amounts to a leap of faith and make some sort of assumption that this or that animal is 'like us' in having brain structures or behavior that leads it to suffer like us." There are different ways to bridge this gap (or not): some argue for similarity of brain structures; some argue from continuity in behaviors. It is worth remembering that scientists still don't understand human consciousness and still puzzle over what makes a human being different from an extremely smart computer or how exactly the electrical impulses in our nervous system give rise to "conscious experience," to feelings such as love, shame, or pain. Yet we don't deny consciousness in humans, despite its philosophical and scientific mysteriousness. We happily jump over the explanatory gap when it comes to other humans.

Because Dawkins is skeptical about animal emotion—we cannot *know* that animals experience subjective emotional states at all—she thinks that it "requires a leap of faith" to apply the word "suffering" to nonhuman animals. She may be right. And she admits that she is willing, though reluctantly so, to make this leap. The leap, for me, is both infinitesimal and obligatory. It would require a much more precipitous leap of faith for me to *deny* that animals suffer.

Jaak Panksepp, a neurobiologist who studies the physiological roots of animal emotion, believes that we can *know* that animals have emotions by studying their brains and observing and measuring physiological events in the brain, such as changes in the levels of certain chemicals. He provides considerable evidence that the same basic emotional systems are present in all mammals: seeking, rage, fear, lust, care, panic, play. Other affect systems that are not properly emotional are also present: pain, pleasure, disgust, hunger, and thirst. If these emotional systems are in place, don't we then have the necessary raw material for profound suffering? And can we dispense with all this leaping over philosophical crevasses?

ANIMAL PAIN/HUMAN PAIN

Pain in animals can be assessed using two basic tools, which are intertwined. First, there are objective scoring systems based on measures of clinical status such as respiration and pulse, combined with behavioral cues that signal when an animal is in pain. Behavioral and physiological responses of animals to pain are similar across mammalian species. We see an increase in heart rate and respiration and a disruption of normal behavior. An animal in pain will typically move less, eat less, and be socially withdrawn. The pain detection threshold—the point at which pain is first perceived during noxious stimulation—is essentially the same in humans and other warm-blooded vertebrate animals. Long-term psychological responses to chronic pain are also similar across species: animals become depressed and experience chronic anxiety.

Second, scientists use what is called the pain-equivalence test. You assume that pain in human and nonhuman animals is essentially equivalent and you ask yourself, "would this hurt me?" If the answer is yes, then the presumption is that it will also cause pain to an animal. So, if a medical intervention or disease process is painful for us, we should assume that it is also painful for other animals. It is worth reminding ourselves (over and over) that much of what we know about human pain—types of pain, modalities of treatment, effectiveness of various analgesics, side effects of medicines, even psychological nuances of the pain experience—comes from research on pain in animals. For the sake of science, their pain is like our pain.

Still, there are many challenges in accurately assessing animal pain, and as much as their pain is like ours, it is also probably *not* like ours. For example, although the pain *detection* threshold is similar across species, the pain *tolerance* threshold—the greatest level of pain a subject is prepared to tolerate—is not. This can vary quite a bit by species, and even within a given species there is considerable variation. Research has shown, for example, that pain tolerance among humans can be moderated by age, gender, experience, cultural attitudes, meditation, and "priming" (the power of suggestion).

Inferring how pain feels to an animal with reference to our own experience works especially well for the animals most similar to us, but it becomes more challenging when the species under consideration is quite different in physiology and neural circuitry, like the brown trout. When it comes to ticks, spiders, and other invertebrates, we simply don't know enough about their behavior to interpret potential signs of pain, and assumptions about equivalence may lead us to assume wrongly that pain cannot be experienced by creatures so unlike us.

The phrase "translational pain medicine" refers to the translation of research, usually on rodents, into human clinical practices. I've said that animal research has provided much of what we know about human pain and its treatment. But it is worth noting that pain researchers are themselves frustrated by how poorly the "translation" has been working, and this points to the danger of taking human/animal equivalence too far. Part of the problem is that common pain models don't actually mirror real clinical conditions. For example, one very common pain model is the acetic acid writhing test, in which acetic acid is injected under the skin of a mouse or rat to induce a burning sensation. Yet how many people come to the clinic for treatment of acetic acid under the skin?

Pain researchers further note that many of the behavioral responses used to measure pain in rodents—licking, biting, vocalization—can be seen in "decerebrate" animals (animals whose brainstems have been transected, or cut): the vertebrate spinal cord will react to noxious stimuli, even without an attached brain. In order to study the conscious perception of pain, behavioral measures that involve cortical structures would be necessary. This raises questions about whether behavioral responses are a reliable indicator of what's happening inside the mind of an animal.

Yet another confounding factor in the use of animal models is the individualized nature of pain. Almost all pain research is conducted on a homogenous population of adult male Sprague Dawley rats. ("Sprague Dawley" is a breed of albino rat developed in the 1920s by the Sprague Dawley animal company and used extensively in medical research.) And the best way to get reliable results is by downplaying the possibility for confounding factors when collecting a set of data points. But human beings, of course, are quite different from male Sprague Dawley rats. And female humans react to pain and analgesics differently than do male humans. Race, ethnicity, age, experiential background, and mood all also influence both the perception of pain and the effects of various analgesics. Every individual has a unique reaction to pain, and this individual variation is found among animals as well, even among those lovely male rats.

Finally, we need to give some thought to how differences between human and animal minds may influence suffering caused by pain. Although we might assume that the greater complexity of the human mind means that we will suffer more deeply from pain than other, less complex, animals, this may not be true. Consider the observations that follow—from a scientist, a veterinarian, and a philosopher. Our scientist notes that the adrenal response to stress is more pronounced in animals than it is in people. Why? It may be because people can deal with stressful situations using psychological tools that animals lack—for instance, we can understand why we're being poked with a needle. Our veterinarian suggests that animals may suffer more severely from pain than people do. Pain, he says, is divided into a sensory-discriminative dimension and a motivational-affective dimension, and since animals are more limited in the first dimension, they may have more pronounced reaction in the second dimension. And finally, our philosopher points out, "If animals are indeed . . . inexorably locked into what is happening now, we are all the more obliged to try to relieve their suffering, since they themselves cannot look forward to or anticipate its cessation, or even remember, however dimly, its absence. . . . If they are in pain, their whole universe is pain; there is no horizon; they *are* their pain."

Human pain is not identical to animal pain. But they are similar enough that we can use our understanding of human pain to make cautious inferences about what animals are feeling. Empathy is the art of perceptive inference—"intuiting" the feelings and needs of another and responding appropriately. We do this with each other all the time, and we can do it with animals as well.

THE PAIN PARADOX

Animal pain presents a paradox. On the one hand, scientific research on animals has proceeded under the assumption that animals do not feel pain. Animals may have pain receptors, but since they don't have emotions, they don't *feel* pain. On the other hand, research into the physiology and even the psychology of human pain has traditionally been conducted using animal models, on the scientific assumption that animal pain and human pain are essentially equivalent.

One of the people most instrumental in trying to call science to account for this paradox is Bernard Rollin, a professor of bioethics and philosophy at Colorado State University. Dr. Rollin was involved in the establishment, during the 1980s, of animal care and use committees at research institutions and has devoted his entire career to working with veterinarians, researchers, cattle ranchers, and pig farmers to improve animal welfare. During the early 1990s, when Dr. Rollin wrote *The Unheeded Cry*, it was still commonplace to deny emotions in animals. Yet in the two decades since, research into animal cognition has proven beyond any reasonable doubt that animals do, in fact, have consciousness, experience a whole range of complex emotions, and most definitely have cognitive apparatus necessary to perceive pain.

I wanted to find out how this more advanced science of animal cognition has influenced attitudes toward animal pain, particularly among veterinarians. Since Dr. Rollin works closely with veterinarians and veterinary students, I thought he might be in a good position to assess how things have changed. Colorado State isn't far from where I live, so I decided to pay him a visit.

We met on a brisk winter morning in the philosophy department at Colorado State. When I arrived he was leaning on a table in the hall, talking on his cell phone. I had set up the interview by phone and had never met him in person, but he was easy to identify by his deep voice and clipped speech and, especially, by his liberal use of cuss words. Although he is short— probably five foot six, tops—he has "presence." He seemed to fill up the conference room where we sat down talk. His chest is enormous, and his graying beard fans out from his face, making him look at bit like Karl Marx. One of the first things he told me is that he rides a Harley—and he looks the part, with black boots and black jacket—and, as an aside, he said that he benches five hundred pounds. "People just don't fuck with me," he added. And even though he could ram his philosophical ideas down your throat, there is a gentleness to him.

I asked him about how attitudes toward pain management have changed in the past decade or two among the veterinarians and students he works with. He let out a sigh and scratched his head, then ran his hand through the graying hair. "They are barbarians," he said to me. "Really, they are."

I'm sure the surprise registered on my face.

"The current wave of vet students, they are worse than ever. It is so discouraging. I'm not a chauvinist," he went on, leaning forward in his chair and looking hard at me. "If you think I am, then so be it. But I'm not. But I have to tell you, 92 percent of the vet students now are female. And things are getting worse. You're swimming against the tide if you try to be compassionate."

When I asked him why science was so slow to acknowledge animal pain, he said, "There are layers upon layers of ideology, that's why." Every one of his vet students, by the time they graduate, will have come into contact with someone—a professor, a fellow student, a practicing vet—who doesn't believe that animals feel pain. And some of them will, themselves, still ignore animal pain. It's just how they are taught to think.

Early in his career, Dr. Rollin said, scientists rarely acknowledged that animal pain was real and when they did, they conceptualized it in purely mechanistic terms. Even during the sixties, vets were not trained to take animal pain seriously. Students were taught to castrate horses using a curariform (muscle-relaxing drug) such as succinylcholine chloride. This class of drugs causes neuromuscular collapse but not loss of consciousness. He quotes a dean of a vet school saying, "Anesthesia and analgesia have nothing to do with pain. They are methods of chemical restraint." (Stop for a moment and think about surgery under "chemical restraint"—that is, where you would be essentially paralyzed but still aware of the pain and the conduct of the procedure. Then think about your worst nightmare. Compare the two.)

Scientists, he said, have tried to smooth over the paradox of animal pain by treating pain as a mechanical, physiological event (as "nociception") rather than a mental state or mode of awareness. They talk about pain *responses* and dismiss questions about animal *feelings*. Something similar happens with the concept of stress, which is used as a catchall for fear, anxiety, and any other sort of misery. One can thus say that noxious situations have deleterious effects on animals without invoking awareness or consciousness (or excessive empathy).

Dr. Rollin seems haunted by his past, by what he has seen happening on the front lines of a veterinary medical school, and like a veteran of war, he recalls one horror story after another. "You know about the multiple surgery thing, right?" he asked me at one point. As I shook my head, he went on to tell me about a time, earlier in his career, when several of his vet students came to him to report a problem. The students learned to do surgery on animals by—of course—practicing on animals. In one surgery class, the students were practicing on dogs. Each student had a dog, and on each of these dogs (about 120 of them for one surgery class), the student would perform nine surgeries. After the ninth surgery the dog would be killed. This was at a time, he reminded me, when it was common practice to perform surgeries without anesthetic or anesthesia. "There was no pain control for these animals," he said. Between each surgery, the dog would be given a bit of time to heal, and then it would be put under the knife again.

"One kid told me," he continued, "that in the middle of a surgery the professor decided it was time for lunch. So they just left the dogs there on the tables and went for lunch." I watched this Harley-riding man's eyes fill with tears.

Old attitudes toward animal pain are yielding to a new approach. Unlike Dr. Rollin, who is frustrated by what he sees happening now (or not happening), my impression is that things are improving. A decade ago, Dr. Sheilah A. Robertson, opening speaker at the 2001 American Veterinary Medical Association's Animal Welfare Forum announced, "The scientific evidence is overwhelming that animals do feel pain. What we need to do is now move on and discuss the more important topics like how can we help them."

One of the people working hardest to improve our response to animal pain is Robin Downing, a veterinarian practicing in Windsor, Colorado. She attended veterinary school in the 1980s, when the dogma was very much as Bernie Rollin described it: animal pain, when acknowledged at all, was viewed as an important form of restraint—pain keeps injured or ill animals quiet. Dr. Downing knows exactly when things turned for her. She was asked to treat a red heeler with a bowel obstruction. She had three options: euthanize the dog, do the surgery without pain medication because nothing was available (knowing full well that the pain of surgery would probably kill the dog), or find a new way. She chose option three. She consulted with a human doctor who specialized in pain management and used what she learned to anesthetize the heeler and perform a successful surgery. Since then, she has continued collaborating with human pain specialists and has put animal pain management on the map. She opened the first veterinary pain management clinic and founded the International Veterinary Academy of Pain Management.

Still, Dr. Downing concedes that the landscape of animal pain remains very bleak. The majority of animal pain is undertreated, mistreated, or not treated at all, and a great deal of unnecessary pain exists. Routine pain management is not practiced in the majority of veterinary settings, and many dogs and cats are euthanized simply because pet owners know of no other option. Most veterinarians are woefully untrained in understanding and treating animal pain, and so even though a great many options are now available, they are put to little use.

PAIN TREATMENT AS A BASIC FREEDOM

Pain is not an unalloyed evil; it serves an important evolutionary function and keeps animals safe and alive. Yet untreated pain—pain that could be managed with medications or alleviated through closer attention to the needs of an animal or which is deliberately inflicted on an animal—is something the world would be better off without. If we can manage pain effectively, Downing notes, we'll see increased survival and better quality of life for our animal companions.

Within the human realm, access to pain treatment has been framed as a basic human right. Human Rights Watch published a report in 2009 titled "Please, Do Not Make Us Suffer Any More . . . : Access to Pain Treatment as a Human Right." The World Health Organization, as the report tells us, estimates that about 80 percent of the world's population has "either no or insufficient access to treatment for moderate to severe pain" and that tens of millions of people each year suffer from untreated pain at the end of their lives. The organization speaks of untreated pain as a major public health crisis. This is particularly true for people in developing countries. Under international law, countries are obligated to make available adequate pain medication. Failure to provide pain medication is both a violation of the right to health and, in some cases, a violation of the prohibition against cruel, inhumane treatment. At the very least, says the report, all nations must ensure the availability of morphine, which is considered "an essential medicine that should be available to all persons who need it."

Untreated pain could be considered a canine and feline public health crisis, too. Attention to animal pain is a basic moral obligation for those who own or treat animals, and failure to treat pain is cruel and inhumane. I don't particularly like the use of rights language in relation to animal welfare; instead, I prefer the language of our Six Freedoms. Number three of the Brambell Commission's original freedoms is "Freedom from pain, injury, disease." Ensuring this freedom means making pain management a basic tenet of caring for our animals.

This freedom is far from adequately realized. Too many animals—even well-cared-for and beloved companion animals—suffer from pain. This is sometimes due to neglect or outright cruelty on the part of owners and sometimes due to inattention by veterinarians. But most often, it is simply that pet owners don't know that their animal is suffering or don't understand how to manage pain effectively (I myself have failed in this regard), and veterinarians, for whatever reason, don't do as much as they could to alleviate animal pain.

It is important to get a handle on pain because it figures so significantly in decisions about aging, dying, and death in our pets. For better or worse, pain is the fulcrum around which a great many end-of-life decisions turn. How, if at all, should pain guide our decisions about whether to euthanize an animal? If pain in animals could be adequately managed, would hospice care and natural death be a better alternative to euthanasia? To phrase this another way, do we euthanize our animals because we lack more effective treatments for pain?

ROOM FOR IMPROVEMENT

As Rollin and Downing both note, there was a time not too long ago when pain medications were not used in veterinary medicine. Operations were performed without anesthetic, and chronic pain associated with cancer or osteoarthritis was essentially ignored. While this rarely happens today, careful pain management has nonetheless lagged behind other areas of veterinary medicine. Given what we do know about animal pain, it is surprising to see that pain treatments are not standardized or consistent and that so little attention is paid to effectiveness.

A survey at Colorado State University School of Veterinary Medicine found that most vets agree that animals experience pain much as people do, but they disagree widely about when to treat pain. And, in fact, some veterinary surgeons still believe that pain is necessary to keep an animal quiet following surgery and deliberately undertreat pain. Male vets and vets who graduated more than ten years ago are, according to the data, less likely to treat pain. Veterinary ethologist Kevin Stafford, in *The Welfare of Dogs*, gives some other disquieting statistics. He writes that in Britain 93 percent and in Canada 84 percent of veterinarians used analgesics for orthopedic surgery. This means that between 7 percent and 16 percent of all orthopedic surgeries were conducted *without* any analgesia. He goes on: "Veterinary use of analgesics appeared to be influenced by assessment of the painfulness of the procedure, and they were used by 68% for thoracotomy, 60% for cruciate ligament surgery, 53% for lateral ear resection, 34% for mastectomy, 32% for dentistry, 29% for perineal hernia repair, 22% for toe amputation, but only 6% for ovariohysterectomy and 4% for castration."

Veterinary practices are shaped by ongoing veterinary research, and a great deal of this research focuses on assessing the relative painfulness of various procedures and of various pain management tools. Yet the tools may be too blunt for the fine work they are asked to perform. For example, based on "pain scores derived from behavior-based composite pain scales" we would be led to assume that there is no difference in the pain experienced by dogs undergoing ovariohysterectomy compared with those undergoing ovariectomy, even though incision lengths were significantly greater in dogs that underwent ovariohysterectomy, compared with those of dogs that underwent ovariectomy, and even though ovariohysterectomy involved removal of ovaries *and* uterus, not just ovaries. It truly could be that there is no difference in the painfulness of the two procedures; it could also be that pain scoring systems are mighty crude and that they obscure as much (or more?) than they illuminate. Even with the administration of analgesics to dogs postoperatively, we cannot be confident that they receive effective pain relief because veterinarians lack information about effective dosage rates and duration of effectiveness.

Effective pain management rests not only on clear veterinary understanding of animal pain but also on the willingness and capacity of pet owners to be careful guardians of their animals. At my local humane society, provision of pain medication following spay or neuter surgery is left up to the animal's owner. I am not a cat or dog, but I can say with confidence that if you cut out my ovaries, I would most certainly want some vicodin or codeine—and maybe a couple shots of whiskey. But from what volunteers there tell me, many people forgo the extra fifteen dollar charge. Perhaps the fact that the pain pills are presented as optional, and at additional cost, gives the impression that they are not strictly necessary but, rather, an extravagance?

As with humans, it may be that chronic pain in animals represents an even more serious trouble spot than does acute pain. A great deal of chronic pain goes untreated. Chronic pain is more difficult to recognize than acute pain—the signs are more subtle, and there is no obvious physical insult to clue us in to the animal's pain. Chronic pain often develops gradually (as in the case of pain from osteoarthritis), so any behavioral changes may also take place gradually—and thus may fall beneath the radar of our attention, even if we are taking good care of our companions. Furthermore, many behavioral changes, such as a gradual slowdown, may be attributed to aging, while the underlying (painful) disease process goes untreated. Osteoarthritis is one of the most common joint diseases in dogs and a common cause of chronic pain. Stafford estimates that some 10 million dogs in the United States suffer from osteoarthritis at any one time and that only a small number of these are actually treated. Of those who are treated, he estimates that many will be treated ineffectively, with too little pain medicine over too short a time span.

Yet another common scenario is the well-meaning pet owner who does her best for her animal but winds up doing more harm than good. For example, a recent posting on the *New York Times* blog *Well Pets* taught me something I didn't know: ibuprofen can be poisonous to dogs. The story reminds us that what might help our pain might not, in parallel ways, help our pets. The blogger was trying to alleviate the discomfort of her German shepherd, who has an arthritic leg. Being a physician, she figured that high-dose ibuprofen would help. After several doses of the ibuprofen, the dog stopped eating and lost control of his bladder. Our kind blogger called the vet, who told her to take the dog to an animal hospital immediately. It turns out that ibuprofen can cause ulcers, intestinal bleeding, kidney damage, and even renal failure (i.e., death). (The German shepherd was okay— but only after a $3,000 veterinary bill.)

HOW DO WE KNOW THAT AN ANIMAL IS IN PAIN?

Good question and, unfortunately, one that is hard to answer with precision. In fact, one of the most difficult aspects of treating pain in animals is knowing when pain is present. In human pain medicine, the guiding rule is "Pain is what the patient says hurts." Over and over, we are reminded of the subjective nature of pain. Even the same stimulus can have vastly different effects on different individuals. Some people hardly notice a simple flu shot; others find the injection quite painful. But animal patients can't tell us with words what hurts.

We can measure physiological events such as increased respiration and pulse, which provide clues to distress, but these won't tell us all that much about how an animal is feeling. The National Research Council concludes that "there are no generally accepted objective criteria for assessing the degree of pain that an animal is experiencing." Robin Downing believes that one of the most important challenges facing veterinarians is the lack of a simple, objective, consistent tool with which to measure animal pain.

Because of their training, veterinarians are in the best position to read the behavioral signs of pain, but veterinarians spend very little time with individual animals. In a fifteen-minute office visit, they can only read certain cues, especially with an animal that they may not know very well. Yet we companions, who know our animals the best, are unskilled, untrained, perhaps unobservant. I have no doubt that I have overlooked pain in Ody simply because I can't "read" his behavior in a sophisticated way and don't really know what to look for.

Many animals in pain change their normal behaviors or even physical appearance. Signs of pain may be obvious, such as vocalization, writhing, and struggling. But often the signs are more subtle: changes in the autonomic nervous system (salivation, dilated pupils, increased heart rate), panting, increased body temperature, shivering, and piloerection (i.e., goosebumps). The International Veterinary Academy for Pain Management lists the following as signs of pain to look for in our companion animals: posture (tucked abdomen, drooping head, arched back), temperament change (aggression, avoidance of social interaction), vocalization (one of the least common signs of pain), movement (reluctance to move, prolonged lying or sitting, lameness), decrease in appetite, and a decrease in grooming. The factsheet "How to Recognize Pain in Your Dog" says to trust your instincts. If you think your dog is in pain—if the question even crosses your mind—he or she probably is.

Of course, being able to detect changes in normal behavior means knowing how normal behavior looks. I may know Ody inside and out, but I can't say I know everything there is to know about Ody or, more generally, about dog behavior. And sometimes—most confounding!—one of the behavioral signs of pain is to show no pain. "Stoicism" occurs in many species, most notably in cats. The survival value of stoicism is obvious: you don't want predators to know you're hurting, or you'll become a target.

The behavioral repertoires of animals vary by individual and by species. To read the signals of a given animal, you have to know their particular language. Rats' coats get puffy and dull. Horses will tighten certain muscles in their eyelids. Veterinarians used to be taught to read humanlike behavioral signals as evidence of pain in animals. But this technique didn't work so well. For example, after surgery a cow will immediately begin to eat. Humans in pain don't eat, the reasoning goes, so cows who are eating must not feel pain. But in ruminants, eating despite pain would confer evolutionary benefits: a cow who didn't eat not only would become weak but would also appear weak to predators (a cow not grazing with the rest of the herd would be an anomaly). Similarly, it was assumed that dogs didn't feel pain because they would be up and moving about directly after abdominal surgery. But the musculature of the abdomen is different in dogs than in people: in dogs, the abdominal muscles do not hold the body erect (the viscera in dogs are suspended from a mesentery "sling"). Similar accounts can be given for many other instances in which animals behave differently from humans when in pain.

Elderly animals probably suffer the most from untreated pain. Many people—myself included—take a passive approach to aging in their pets. We wait until signs of pain or discomfort or disability are loud and clear, and only then do we venture to the vet. At this point, an animal's medical problems may be quite advanced and more difficult and painful to treat. Even when signs of pain or illness are obvious, many pet owners neglect to seek medical attention for their animals, attributing troubles to "just getting old."

Is Ody in pain? I've wondered this many times as he ages and becomes increasingly disabled. And even though I've been steeped in the pain literature for the past year, I am still uncertain. My best guess is: a little, but not too much. He is stiff and sometimes bites at me if I touch his lower back. But he is contrary, and he tries to bite me if I pester him to leave the kitchen while I'm cooking. He has been panting for the past two years, pretty much nonstop, but this, I think, is from his laryngeal paralysis. I've tried some of the arthritis medicines like Rimadyl, but as far as I could see, nothing changed. Tramadol, a pain medication, also seemed to have no effect. My vet tells me that because Ody's lameness is due to neurological deficits, he probably feels no pain. But I'm not sure I believe this. Ody is, as always, inscrutable.

I have noticed, oddly enough, that Ody's end of life feels considerably more complicated precisely because he appears not to be in pain. If he were in pain, I would try to alleviate his pain through treatment, if possible, and at some point the pain would become severe enough that we'd euthanize him and put him out of his misery. But without pain as a weight on the scales, what do I use to measure his distress? When he gets stuck in a corner behind the leg of the piano and can't stand up because his rear legs are too weak, and I find him lying in his own feces? When he falls over several times a day, and flails around like an upended beetle until I find him and lift him upright? When he cannot lean his head down to his dog bowl without toppling over? When he wears a look of worry all the time? How do these figure on the scales of suffering, particularly when otherwise he seems very much intact and very much himself—always hungry, always under foot, always the big red dog?

TREATMENT OF PAIN

One of the most common "treatments" for animal pain is euthanasia. And sometimes, perhaps, death is, in fact, an appropriate choice. But most of the time, there are much less draconian ways to manage animal pain. The list of possible treatment options looks similar to the human pain armamentarium: local anesthetics (lidocaine), steroids (prednisone), opioids (morphine, tramadol), and nonsteroidal anti-inflammatory drugs (carprofen, meloxicam). Many nondrug therapies also help with pain in animals, including heat, ice, massage, nutrition, physical therapy and exercise, acupuncture, homeopathy, and neutraceuticals such as glucosamine chondroitin.

Despite all these options, treating animal pain is complicated. Human pain treatments cannot just be transferred to animals (or vice versa). For example, the analgesic drug meperidine is dosed, in humans, every four hours for effective pain management. So, it was also given to dogs and cats every four hours—until further study revealed that the effectiveness of the drug in animals often peaks at two hours, after which it provides no pain relief.

Pain treatment also has to be aimed at the right kind of pain. Origins of pain can be somatic (skin, bones, tendons, muscles), visceral (internal organs), and neuropathic (nerves, spinal cord, brain). Different disease processes cause different kinds of pain (visceral pain versus neuropathic pain; chronic versus acute), at different levels of discomfort. Effective pain treatment, then, depends on many things, including origin and type of pain, species, individual pain tolerance, and background health status.

Furthermore, pain medications can have side effects, and these always have to be balanced against pain control. Analgesic drug dosing is subject to considerable individual variability, and follow-up care is often needed to balance adequate pain relief with tolerable side effects. Side effects can vary considerably from one species to the next (negative side effects of nonsteroidal anti-inflammatory drugs, such as renal failure, are much more common in cats than dogs), from one breed to another, and even from one individual to the next.

The International Veterinary Academy for Pain Management recommends two basic pain treatment strategies. First is the use of multiples therapies (combinations of different drugs, or drugs combined with massage, physical therapy, etc.) to control pain. Using several different therapies can produce a synergistic response and is generally more effective than any single drug or therapy. Second is what they call preemptive therapy. The use of analgesics prior to the onset of pain will dampen the pain response. Once pain has fully "kicked in" it is much harder to treat. This is why your doctor will often tell you to take ibuprofen an hour or two before coming in for a minor procedure that might cause some swelling and pain. Robin Downing uses the image of a pain-management pyramid, with the base layer consisting of combinations of relatively mild pain therapies such as nonopioids, mild opioids, nondrug therapies like acupuncture, and adjuvant agents that target specific receptors in the nervous system and complement the action of other pain medications. As pain escalates, we add new tools and move up the pyramid to increasingly powerful drugs and higher dosages. The goal is to stay ahead of the pain. At some point, we may reach a level at which euthanasia is the only effective treatment left in our armamentarium.

The upshot here is that even in the hands of a highly skilled veterinarian, pain management is not a simple affair. Even if you've paid close attention to your animal and accurately interpreted behavioral signs of pain, you can rarely just head to the vet, grab a bottle of pills, and be done with it. To treat pain right, you may need to try various drugs or combinations of drugs, in tandem with other pain management techniques, all balanced against side effects of drugs and the stresses of treatment. In the meantime, each of the drugs and the laser therapy, the underwater treadmill, the special food, the acupuncturist—each of these takes time and costs money. Unfortunately, very few pet owners may have the knowledge base, persistence, financial resources, and time to do all that is needed to optimize pain management in a seriously ill or injured animal. And this is why so many pet owners wind up feeling deeply ambivalent about caring for an animal in pain: it is extremely hard to do all that we might do, and even when we try hard, we may fall short of keeping our animal comfortable. This is when we begin to think that euthanasia may be the best way out, for everyone involved.

ON THE OTHER SIDE OF THE SCALE (IS ODY HAPPY?)

When it comes to ending the life of a companion animal, we often focus on their pain: how heavy is their level of pain, relative to their pleasure? In measuring quality of life, we place pain on one side of the imaginary scales and pleasure on the other, with pain perhaps weighing slightly heavier because of its moral urgency. Just as "pain" is shorthand for a whole range of sufferings, "pleasure" is shorthand for positive well-being, including pleasure, happiness, joy, and contentment. Are we any better at seeing the signs of pleasure in our animals? Is Ody happy?

Animals most certainly experience pleasure, as biologist Jonathan Balcombe has carefully argued in three published books on the subject. They may even experience unique forms of pleasure inaccessible to humans. Birds don't just sing out of necessity but actually sing because it gives them joy. Male songbirds have increased dopamine levels when they sing, especially when they sing to a female. Rats laugh when they play with each other and when they are tickled on the stomach by a familiar human handler.

Topaz is joy in flight when he's leaping into the air to catch a Frisbee. When he brings the Frisbee back, he flings it at my feet, looks at me with bright eyes and dancing feet, all smiles. And when it snows—oh, what ecstasy! Topaz will bound up next to me, look at me with a twinkle in his eye and touch his nose to my fingers, and then burst off after Maya. Then he will stop short, as if he's had a sudden inspiration and he'll take off in a new direction, his nose down in the snow, like a miniature snow plow, until his snout and face are completely white. He will launch himself into a big snowdrift and completely disappear under snow. He'll crawl out, shake off, and do it again. Happiness is Maya stalking a bird on our morning walk or watching squirrel TV from the back porch or rolling on her back in the fresh-cut grass at the park; it is Ody standing in the middle of a pack of children or eating a stick of butter left to soften on the counter.

You'll see that I've conflated pleasure with the less scientific-sounding happiness. Indeed, the term "happiness" is rarely applied to animals in scientific contexts. The reason for this, veterinarian Frank McMillan speculates, is the commonly held belief that animals live only in the moment and experience only momentary emotions; they do not cognitively evaluate the arc of their lives as a whole. Animals are capable of momentary pleasures, small happinesses, but not true happiness, the way humans are. McMillan challenges this dogma (pun intended). He believes that animals can also experience "true" happiness—that is, having a pervasive sense over time that all is well.

Humans have what has been called an emotional set point—a baseline level of happiness. Although life's events may temporarily cause a spike or a dip, people tend to return to their baseline happiness level within a few months' time. People who win the lottery or receive a coveted award may experience elation, but before long the thrill is gone and they are back to where they were, in terms of happiness. Likewise, people who suffer profound injury (becoming paralyzed or blind) or a profound loss (the death of a spouse) usually return to their baseline happiness within a few months. According to McMillan, ample data suggest that for animals, too, emotional ups and downs are transitory. Animals have a baseline emotional stability that transcends fluctuations in mood. In one study, the mental attitude of dogs who became paralyzed in their hind legs was just as good as prior to their paralysis, as judged by their owners. Ody's set point is relatively low, compared to happy-go-lucky Maya and intensely happy Topaz. After the thrill of stealing an entire package of hotdogs off the counter has worn off, Ody returns to his same old melancholy self.

Life satisfaction, as McMillan notes, is a reflective appraisal of one's life as a whole. Animals may not make cognitive appraisals, as we do. But many of the factors that contribute to human life satisfaction are relevant to animals: an active engagement with the world, a stimulating environment, the capacity to fulfill one's needs and goals, a sense of accomplishment, and a feeling of control.

ANIMAL SUFFERING AND THE FOUR QUADRANTS

A recent issue of the *Proceedings of the Royal Society B* (*B* for "biological sciences") included an essay titled "An Integrative and Functional Framework for the Study of Animal Emotion and Mood," by Michael Mendl, Oliver Burman, and Elizabeth Paul. Their essay offers us a window into the state-of-the-art on animal emotion. It is revealing both for what it tells us and for what does not, and it sheds some additional light on the question of animal pleasure.

Most research on animal emotion has focused on discrete emotions: fear, anxiety, happiness. But there has also been an attempt to offer an overarching framework for understanding emotion—for answering questions such as "what exactly are emotions?" and "out of what inner morass do they arise?" The authors present their view of this overarching framework, and they offer a nice, simplified visual of emotion.

The visual, which reminds me of seventh-grade math, is a basic mapping of emotion into two-dimensional space, resulting in a four-quadrant grid. This grid represents what the researchers call our core affect system. The two dimensions represented are valence (on the *x* axis) and level of arousal (*y* axis). Discrete emotional states such as fear, sadness, or joy are "locations" in this core affective space (see fig 1). ("Affect" is a term used in psychology to mean emotion or feeling. "Valence" is a measure of the at-

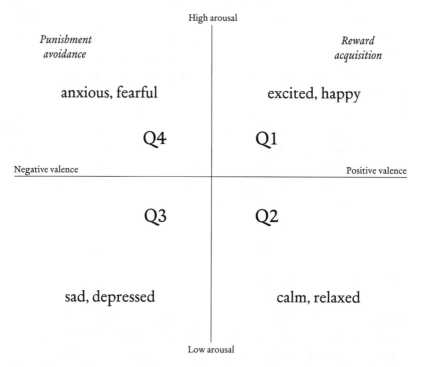

Fig. 1. Core affect system. The two dimensions represented are valence (on the *x* axis) and level of arousal (*y* axis). Discrete emotional states such as fear, sadness, or joy are "locations" in this core affective space. Adapted from Michael Mendl, Oliver H. P. Burman, and Elizabeth S. Paul, "An Integrative and Functional Framework for the Study of Animal Emotion and Mood," *Proceedings of the Royal Society B* 277 (2010): 2895–2904.

tractiveness [positive valence] or aversiveness [negative valence] of a given stimulus.)

Emotion is a biological system designed to motivate an organism to seek and obtain rewards (food, sex, shelter) and avoid punishments (heat, cold, attacks by a predator). Emotions that map onto the right side of the graph—quadrants 1 and 2—are associated with reward: happy, excited, calm. Emotions that help avoid punishment map onto quadrants 3 and 4: fear, anxiety, sadness. The x-axis, representing valence, moves from strongly negative (on the far left of Q3 and Q4) to strongly positive (far right of Q1 and Q2), while the y-axis, representing level of arousal, moves from low (Q2 and Q3) to high (Q1 and Q4). This "dimensional" theory of emotion has gained prominence within research on human emotion. Mendl, Burman, and Paul argue that animal emotions are similar enough to human emotions that this model is fitting for both.

Like people, animals don't experience just one emotion at a time and then move on to another. Rather, animals have mood states. There is a swirling mass of emotions, with some coming briefly into focus and then receding into the background, like images in a crystal ball. Core affective states occur in response to specific stimuli. But they also exist, researchers note, as a kind of background existential state and occur in the absence of specific stimuli. They call these free-floating moods. Mood states are longer-lasting tendencies toward, for example, happiness or mania, sadness or depression; these are an animal's emotional set point. "For example," the researchers say, "if an animal is in an environment in which it experiences frequent threatening events, and hence its emotional state is often in the Q4 quadrant, it may develop a longer-term high-arousal negative mood state that mirrors this cumulative experience. If it is frequently successful at avoiding these events, or it is in a generally safe environment, a longer-term low-arousal positive mood state (Q2: 'relaxed'/'calm') may result."

So, how I feel at the moment places me somewhere in "core affect space." Right now is a Q2 kind of moment, though I may ease up into Q1 in a few minutes when I get up to fetch a chunk of dark chocolate. Ody seems to have a mood state that places him somewhere in Q3. He is hounded by chronic low-level anxiety. His cup is half empty. I can see it in the way he crinkles the skin on his forehead. And, especially, in this: before each walk, he will cautiously approach the door. Then, before actually placing either foot outside, he lifts his nose and sniffs. I have no idea what he is looking for, but sometimes he finds IT, because he will turn around and retreat back inside, or else just stand in the doorway, refusing to budge. If his refusal seems ambivalent, I slap on a leash and pull him out the door. And he gives me a look that seems to say, "You are the worst dog owner in the world."

As we know from self-inspection, as well as from a wealth of psychological studies, mood states influence decision making and cognitive appraisals. For example, people in a bad mood tend to judge ambiguous stimuli negatively, while happy people tend toward an optimistic bias. What this suggests is that although we cannot know with certainty what emotional states animals experience, we can infer their underlying mood by watching how they make decisions. An unfamiliar sound, to a happy animal, would be an invitation to investigate, while the same sound might make a nervous, depressed animal flinch or flee. Research on sheep, rats, dogs, and pigs confirms that mood state influences decision making. For example, a team from Newcastle found that pigs living in enriched environments were more optimistic than their bored counterparts. They were more likely to approach an ambiguous noise to see whether they might get a treat. A study by Michael Mendl (author of our mood-state essay) on optimism and pessimism in dogs found that dogs exhibiting severe anxiety when left home alone have a more negative underlying mood state than dogs who are fine when left alone. This confirms my suspicions about Ody. He is just a Q3 kind of dog.

Emotion researchers disagree about how discrete emotions and the core affect system relate. Some believe in a kind of top-down model in which the core affect state (valence and arousal) gives rise, when combined with appraisals of the current environmental stimuli, to specific emotional states such as fear or excitement. Others argue that discrete emotions come first and that core affect is a kind of cognitive distillation of the experience of specific emotions.

Mostly, this doesn't really matter for my purposes here. Except a couple of things. First of all, it is striking that the same basic model should be used for studying human emotion and animal emotion. It may seem quite obvious that humans *are* animals. But this really represents a sea change in the way people view nonhuman animals, especially within science. Remember that, even a decade ago, most veterinarians and scientists, as well as a good many laypeople, were skeptical that animals feel pain or have emotions, much less consciousness. This is not to say that there aren't still many people who deny that animals feel things—but such denial is really a philosophical stance (a rationalization) more than scientific credo.

Second, it is exciting to see some real progress in the study of animal emotion. Even if our access to the subjective experience of emotion in animals is limited—we don't understand their language very well—these subjective experiences are accompanied by measurable neural, behavioral, physiological, and cognitive changes (e.g., changes in brain activity, heart rate, and cortisol levels). The argument that animal minds are too private, and thus can be ignored, is obviously bogus.

Third, most research into animal emotion has so far focused on Q4, the quadrant of high arousal in the punishment-avoidance system—in other words, fear and anxiety. A mind-numbing number of studies and models have been devised to first induce these states in animals and then study them. As Mendl, Burman, and Paul note, "There is a whole industry devoted to the study of fear and anxiety, largely based on the development of tests to induce these states." Research into animal emotions in the other dimensions, in contrast, is less well-developed. Scientists have explored the neurobiology of reward systems (which drive high-arousal Q1 emotions), but much less is known about animal happiness and other Q2 states or about Q3 states such as animal depression. Having a more robust and well-rounded understanding of animal emotion will certainly allow us treat them better.

Fourth, pain is not a mood state but it interrelates with mood states. Pain can feel worse to someone in a negative mood, and chronic or severe pain can create negative feelings and maybe even, over time, lead to a chronic negative mood state. How pain and emotion interrelate will vary a lot by individual, and we need to stay attuned, as caregivers, to the particulars of the animal or animals in our care. The upshot here is that equating pain with suffering isn't all that straightforward. There are many varieties of pain, many varieties of suffering, and many varieties of individual personality profiles experiencing and processing the pain and suffering in unique ways.

Finally: a thought about animal suffering. Many accounts of when death is appropriate for an animal rely on a weighing of pleasure and pain on an imaginary scale and seeing which side is heavier. But this seems to me—although appealing and perhaps inevitable—also treacherous. Emotional states are temporal and constantly shifting, and being in Q3 and Q4 is part of who we are, as animals. We shouldn't be too quick to assume that all Q3 and Q4 emotions in animals are "bad" and thus should be blotted out. I don't think that many people would chose to live life with only positive emotions. Neither should we assume that all negative emotions harm animals in enduring ways. We also, more to the point, tend to assume that being old and near death causes fear and anxiety and sadness, but this is certainly not true for all people and may, likewise, not be true for all animals either. These may be passing emotions of the elderly and ill, yes. But the underlying mood state may still be happy.

I want to be clear that there is a line to draw, somewhere, on suffering. There is intractable, unbearable suffering that has no "silver lining" and leads nowhere near illumination. A shameful number of animals live miserable lives, experiencing emotion only in Q4, like battery chickens (i.e., chickens confined to tiny wired cages in industrial farming) and dogs in abusive homes where they are punished capriciously and where shelter, food, and water are uncertain. Perhaps our goal with animals in our care should be to create the conditions for a life in which the basic fabric is positive but with sorrows and sufferings, fears, and worries creating dark-colored threads woven throughout.

WHO SHOULD DECIDE WHAT MAKES AN ANIMAL HAPPY?

If, in end-of-life decisions, we engage in a kind of pain/pleasure weighing, who should be the judge of what things to put on the scale, in what quantities? With Ody, the pleasure/pain calculus has been so hard to determine. The pain part is hard, ironically, because his disabilities are *not* painful. Rather than being caused by some painful condition like arthritis, the trouble with his hind end is neurological—the right signals are not getting from brain to legs. He probably has little feeling in his legs and feet. When he knuckles it looks horribly painful, but the vet assures me that he probably doesn't feel much. I think his physical condition is distressing to him, but in ways that I find hard to interpret. On the pleasure side he's hard, too, because his life has always been full of anxiety and inner turmoil. This chronic negative affect seems to have gotten worse as he has aged, and now he rarely looks joyful, even when he's obviously experiencing pleasure, as when someone is petting him or when he is licking a greasy pan. So, I can't clearly say he's in pain and that I ought to end his life; nor can I say that he positively experiences moments of pleasure. But even though Ody is in-scrutable to me, I still think that I am the Person who should make judg-ments about his pain/pleasure calculus. I know him best.

It is we who live with animals who are in position, usually, to confirm what's important to them. I am reminded of a story about Finn, a Weimara-ner who belonged to my friend Pansy. When Finn was fourteen, he cut his leg, and Pansy took him to the vet. The vet did some tests and said that he wouldn't be able to stitch Finn's leg because putting him under anesthesia would be dangerous—Finn had a heart condition. The vet proposed doing tests on Finn's heart and perhaps doing heart surgery so that he would then be able to withstand the rigors of getting stitches in his leg. Pansy resisted, thinking that at Finn's age, lots of poking and prodding and spending time alone in a cage at the vet's office was the last thing he would want. The vet became a bit testy. "You don't want Finn to die chasing a rabbit, do you?" he scolded. Without a moment's hesitation Pansy replied, "That's exactly what I want." She grabbed Finn's leash and together they walked out the door.

YOU'VE GOT PERSONALITY

Anyone who lives with a dog or cat knows that animals have personality, the complex of attributes—temperament, behaviors, emotions, ways of thinking—that make each individual unique. "Personality" is an unfortu-nate word choice here. Perhaps "animality" would be better, or "dogality" and "catality." But I'll stick with "personality" since it is familiar to us all, and since this is the term that scientists studying animal personality use.

Over the past decade, the field of animal personality research has blos-somed. Researchers have noticed that the behavioral profile of individual animals within a given species can vary, sometimes dramatically. Not only this, but these individual quirks of temperament seem to remain constant over the lifetime of the animal and from one setting to another. Scientists have found this individual variation across a huge swath of species, from chimpanzees to pumpkinseed sunfish, to barnacle geese, to spiders. Per-haps even more surprising, scientists have found that animal personalities can reliably be tested and described using the same five categories applied to human personality types: neuroticism, conscientiousness, extraversion, openness to experience, and agreeableness.

Ody is a people person through and through—he has no patience for balls and flying discs and no particular interest in other dogs, if there are people around (with the exception of the lovely bloodhound Samantha and the bread-loaf, Daisy). He would rank high on extroversion. I would also rank Ody very high on neuroticism, midway on openness, and low on agreeableness and conscientiousness. (No doubt his self-assessment would be different.) My rating might surprise friends and family who know Ody. I am wrong about agreeableness, they might say. Ody is one of the friendliest dogs you could ever hope to meet. Friendly, yes, but not agreeable. At heart he is contrary.

On the topic of animal personality, evolutionary biologist David Sloan Wilson says, "There are low information processors who don't attend much to their environment and bulldoze through life. Then there are the sensitive ones who are always taking things in, which can be good because information is valuable, but it can also be overwhelming." Studies on highly sensitive people show that their delicacy is "domain general": they are moved by Hallmark TV commercials and bothered by violent movies; they are also sensitive to drugs such as caffeine, and their skin is sensitive to lotions and cleaners. Ody is a sensitive dog—sensitive to the faintest noise that might sound remotely like a gun,which will send him into paroxysms of anxiety, sensitive to whatever dangerous smells lurk outside the door, sensitive to clouds in the sky, sensitive to a my friend Liz's boisterous laughter. Sensitive enough to drive me insane.

Just as animals can be described using traits usually reserved for humans, humans seem drawn to make comparisons to animals. I took a little Internet test to discover "the animal in me" (my spirit totem), and it turns out that the animal whose traits I match most closely is the bat. I am not upset by this comparison. I actually find bats quite intriguing. Like an astrological forecast, I find some truth in my animal description: "Since it is not a true bird and has not mastered the art of smooth controlled flight, it often appears awkward in social situations. But as compensation for this social ungainliness, many bat personalities sport a built-in radar which enables them to intuitively read the motivations of others."

The study of animal personality is important for many reasons, not least of which is that in trying to understand, treat, minimize (and prevent!) animal pain, and to maximize animal pleasure, we must pay attention to the uniqueness of each individual.

PALLIATIVE CARE: PAIN MANAGEMENT AT THE END OF LIFE

Palliate: c1540s, "to alleviate without curing," pp. of L.L. palliare "cover
with a cloak, conceal."

Within human medicine, the treatment of pain is often called palliative
care. Although general palliative care covers all aspects of effective pain
management, a good deal of palliative care takes place at the end of life and
is often coupled with hospice.

Palliation, like hospice, is a philosophy of care. Rather than focus on
curing disease or extending life, it instead focuses on reducing the severity
of symptoms, making patients more comfortable, and improving quality
of life. It is largely about pain control, but it is also bigger than this. It is a
comprehensive and holistic response to physical pain and also to psychological, social, and existential suffering.

In human medicine, palliative care has been relatively slow to develop.
Even thirty years ago, medical students were not taught a great deal about
pain management and were often uncomfortable treating pain. Now there
is an entire subspecialty of medicine in pain management and palliative
care, and though there is still need for considerable improvement, we are
managing pain better. Palliative care for animals has also progressed slowly.
Veterinary medicine seems to be following an arc similar to that of human
medicine (though perhaps two decades behind), with increasing attention
to the nuances of pain management and a more careful response to the
needs of aged, ill, and disabled animals. Euthanasia is no longer the first
line of defense against aging or pain. More people are embracing a philosophy of care that attends to a broad span of animal needs, beyond simply
providing surgery or medicine. And as we'll see in the next chapter, animal
hospice is increasingly popular. Although we have a long way to go until
"dying like an animal" is associated with a peaceful and painless death, we
have made some important strides in this direction.

The Ody Journal

SEPTEMBER 20, 2010

We went to Kansas for two days and left the dogs at Pansy's. When we picked them up, Ody actually seemed happy to see me. He wagged and leaned up against my leg. Usually these days he seems indifferent.

Ody looked noticeably older when we got him home again. His back legs have gotten weaker, and he can't hold himself up when he's standing. Could he have gotten worse in just two days? When he tries to stand still, he immediately begins to sink down in the rear. Three times I saw him sink all the way down, unable to stop gravity's pull. And then his rear legs just kept on slipping, until they were stretched straight out under him awkwardly, and he was stuck, unable to pull himself up. I had to lift his rear each time. Maybe I should be looking into one of those doggie wheelchairs.

He seems famished. I was giving the dogs some Nature's Balance treats—the real meat kind that you slice yourself. Ody seemed desperate to eat as much as he could. He ate chunks of canned food straight off the spoon. Then I hand-fed him some kibble. He ate and ate. You hear about people having to hand-feed their dogs when they get really old; maybe I'll be hand-feeding Ody. The whole time he was vacuuming food from my hand, his legs were sinking down so he was almost but not quite sitting.

SEPTEMBER 22, 2010

Last night Ody was really restless. He barked to go out at about 1:30, as usual. He went out and peed, and then came back in. As usual. But typically, we would all go back to sleep (or he would, and I would climb back into bed and lie there like the insomniac I've become). Last night, though, he began to bark again, about five minutes after he had gone out. I let him out again, thinking maybe he had to poop, too, but had forgotten. But he simply walked outside and then right back in. I heard him bark again, a little while later. Chris got up and tried to let Ody out. But Ody simply stood by the door. What was going on? Is he uncomfortable?

Today I called the new vet, the in-home one that Pansy recommended. She's coming day after tomorrow. I want another opinion about Ody's back legs and whether there is anything we can do to make him more mobile.

SEPTEMBER 27, 2010

Ody seems to be sliding faster now. He sleeps more, gets up less often, and has less interest in walks. For two days he walked only so far as Dale and Melissa's driveway, two houses down; last night, he made it outside into the yard and then turned around and wanted to go back. He is eating less and looks very thin. If I make him homemade food, he'll still eat with interest. And if I feed him canned food by hand, he will eat. But otherwise, he won't. He seems hungry, standing in the kitchen doorway while I make his breakfast. But then he doesn't eat.

The new vet came last Thursday. She is very personable, and the dogs really took to her (even Topaz; remarkably, he didn't bark but just put his ears down in a "hello" and wagged his tail like crazy). The vet watched Ody stand and walk and felt him up and down. Then she lifted each rear leg and placed Ody's foot fur-side down. After a short time, Ody righted the foot. But there was a delay. "See, he can't feel what's happening," she told me. "The signals just aren't getting from his brain back to his feet."

"It's definitely neurological," she said. "The good news is that this means he isn't really in pain. He just can't feel much back there at all." Knowing that the bad news is coming, I sit still at the table. "The bad news is that there isn't much we can do."

The one treatment option we could try, she told me, is a course of steroids. This might help him with mobility and also with the laryngeal paralysis. Given his age, though, she thought we should do a full blood panel before starting him on any medication. Just to be safe.

She needs to take a blood draw, so I stand next to Ody and hold his head so she can find a vein and insert the needle. Given how decrepit he is, Ody shows remarkable strength. It takes about fifteen minutes of struggling for me and the vet to keep Ody still enough to draw one vial of blood. The whole time Ody is panting and quivering and doing his best to bite my arms. The vet seems relieved to be on her way out the door and says she'll call the next day with results.

SEPTEMBER 28, 2010

The vet called and told us that Ody's liver enzymes are dangerously elevated. She thinks it is probably bone or liver cancer but says that it is hard to know for sure. Now we have the dilemma of whether to do more testing. The vet seems eager to do more, but I'm not sure. I tell her we'll think about it. Given how hard the blood draw was for Ody, I hate to think how stressful more aggressive poking and prodding would be. Chris and I wonder together, what will the information give us that we don't now have? We don't plan on any aggressive treatments; there aren't really treatments for bone cancer or liver cancer anyway. Even if there were, we would likely decline, given Ody's age. We could do tests to know what kind of cancer he has. But again, what would it offer? The one advantage would be greater knowledge of what we're dealing with, so that if there were some emergency (internal bleeding?), we would know the reason and wouldn't be searching for a cause, especially in crisis mode. But if Ody had a medical emergency, it would probably just be the end.

Because Ody's liver enzymes are so high, steroids are out because they can really stress the liver. This is disappointing.

The blood panel also tells us that Ody's thyroid is not functioning properly, and the vet recommends Thryosin. We'll start Ody on this tonight.

OCTOBER 3, 2010

I am actually glad to come home and find that Ody has pulled a bag of peanut butter cookies off the counter and shredded the plastic bag all over the floor. He still has some spunk! And not a single crumb left on the floor for me to clean up.

OCTOBER 4, 2010

I would almost say that Ody is perkier on the thyroid medicine. He spends more time awake and seemed, yesterday, quite eager to go for a walk. But his hind end is worse. I just heard a weird noise from the living room—a thumping and thrashing. Ody was stuck on the floor, unable to get up. We've put carpets and runners all over the place because the wood floor is so difficult for him to navigate. But he managed to find one of the few patches of uncovered wood. He's very wobbly now and I always think he's about to topple over (and he is!).

The temperature is perfect now, balanced on the tip of autumn. For some reason, Ody has been able to go outside on his own again—not out the dog door, of course. But out the human door. We're just leaving the sliding door open and the back porch door ajar all night. The mosquitoes are gone for the year, and I'm not worried about psycho killers getting in (not with Topaz on duty), but I do think about mice. It is worth the risk, though. It feels incredibly luxurious not having to get up several times a night.

He seems both better and worse, after a week on the thyroid medicine. He eats less and looks painfully thin. But he also seems a bit more alert.

He has refused to go on walks the past few days. Sometimes he'll follow Topaz and Maya out into the side yard, but then he wants to go back inside. If he walks at all, he'll go only to Anne and Mike's driveway—about twenty feet. Then he just stops and looks and me and waits for me to come back. He gets the tiniest spring in his step on the way home.

I gave him a special breakfast, since he seems not to eat enough and not to have much appetite for dog food. Dumped a can of tuna on top of a bit of kibble. He scarfed it down, but too fast, it turned out. He threw up the whole meal on the living room floor.

I told the vet that we didn't want to do any more tests.

Chris and I went up to the cabin last night while Sage was at an Anti-*Twilight* sleepover. It is definitely getting harder to have Ody there. He did NOT want to get in the car in Longmont; he no longer shows even a flicker of interest in trips. He looks miserable in the back of the Pilot—standing and panting the whole time, cramming himself into the tiniest, most awkward space (to get away from Topaz?). If there is anything loaded in the back with the dogs—like our water jug, or a suitcase—he'll find a way to stand right on top of it, even if there is plenty of floor space.

At about 1:30 a.m. Ody barked to go out. I let him out and he trundled off to pee and then came right back. From this point on, Chris and I were both unable to sleep. Chris finally went out into the living room to try to sleep next to the heater (it is cold!) and found that Ody had been in there pacing around until he finally pooped on the rug and then kneaded it into the rug by pacing on top of it. This solves a mystery for us: Whenever we find poop on the floor in the morning, it doesn't simply look stepped in; it looks as if someone took a butter knife and spread it into the carpet or over the wood. We thought maybe Ody was lying in it, but he doesn't usually have poop on his body . . . it must be the pacing back and forth.

OCTOBER 14, 2010. HOME.

Heard a thump during the night. Ody had fallen in the office—maybe trying to get off the couch?—and couldn't stand up. I had to help him. This was about 2:30. At 4:30, I heard him again, pacing and thumping. Went to make sure he wasn't stuck. He was just standing in the kitchen doorway.

He is becoming thinner and thinner. I have to hand-feed him. He likes this.

OCTOBER 15, 2010

We were all trooping through the kitchen—Sage and a couple of friends and Chris and I—girls all screaming and giggling, excited to go outside and experience the horror of Chris's Halloween decorating extravaganza. Ody had been hanging around the kitchen all night. Topaz was feeding off the girls' excitement and started barking his special frenetic heeler bark and then he suddenly went for Ody's Achilles tendon—his favorite point of attack.

Ody fell down under Topaz's bite. I pulled Topaz off, but Ody had lost his footing and now was sprawled out on the floor, whites of his eyes flashing, desperately trying to get his legs under him. I was trying to help him up but whenever I touched his legs he reached around and tried to bite me. Then, as he flailed on the floor, three greenish clumps of poop squeezed out. As he squirmed, it smeared all over his back.

OCTOBER 19, 2010

Saw neighbor Anne while walking Ody, Maya, and Topaz. She remarked on Ody: "He looks really different than the last time I saw him" (which would have been within the last couple of weeks). He does, too. He looks so skinny—he is the Bone Man.

I figured out one of his problems this morning: I watched him trying to eat, and every time he leans his head down toward his dish his body sways to the side. He has to raise his head to get his balance, so he can never reach his bowl without falling over. No wonder he seems hungry all the time.

I put his dog bowl up on an elevated tray and it seems much better. I kick myself hard for not noticing this sooner.

OCTOBER 22, 2010

Ody stands and walks as if drunk, swaying dangerously to the side. His left hind leg seems almost not to be working at all. He's much worse today.

Got stuck on the couch and when he finally righted himself and jumped off he fell flat onto the floor. How he manages still to jump up on the couches boggles my mind.

The house is covered with carpets runners, everywhere Ody might possibly want to go. Yoga mats, bathroom rugs, welcome mats—anything we can find that will offer some traction. It looks pretty weird.

Ody eats mainly hotdogs these days, refusing most everything else.

He seems restless today. He keeps coming into the doorway of the office, over and over. He just stands there and looks at me. Never into the office . . . only into the doorway.

OCTOBER 24, 2010

Sage asked me tonight, "Mom, what's wrong with Ody's nose?" I went and looked, and it seems to have transformed overnight. It is no longer red with brown age spots. Now it appears a dusky tan. I get down on my knees and right up close to his face. The nose looks all crusty and dark, like cryptobiotic soil or a patch of dark-brown lichen. It feels dry and scaly and rough. It looks altogether not right. Tomorrow I'll call the vet.

I hear Ody in the night. It is still (just barely) warm enough to leave the back door open, so we haven't been needing to get up to let Ody out. But I still can't sleep. I hear him pacing the house, moving phantomlike up and down the hallway outside the bedroom. A loud phantom he is.

5

Animal Hospice

Those of us in the grip of canine love do what we can to make our dogs healthy and happy. We tend them from puppyhood on, training, feeding, walking, creating daily rituals of greeting, play, eating, and sleep. And then we hit a point where it is no longer clear what loving our animal demands of us. All of a sudden or, as with Ody, in bits and pieces over time, we find ourselves less confident about what will make them happy and healthy. And, as it has felt to me, we wander along the lip of a sharp, fog-encrusted cliff, searching for a way down, into the Valley of Death—a way that doesn't involve pushing our animal off of the edge.

Before starting this book, and based on my limited experience, my expectation was that there will come a time when Ody has very little quality of life left, and after an agonizing period of wondering "Is now the time? Is it too soon? Too late?" I will eventually just have to make up my mind. We'll tearfully load Ody into the back of the car and take him to the vet. It will seem strange, to make an appointment for his death—this isn't the way death is supposed to work. The setting will be all wrong: Ody will be nervous—he hates the veterinary office; he will be panting, his eyes will constrict at the corners, and his tail will curl down. The smell will be overpoweringly antiseptic. He'll be laid out on a cold steel table. I'll feel that I am letting Ody down, and I'll be self-conscious about crying in front of the vet and the technician, who will be strangers to me and to Ody. Those who end his life with a quick poke of a blue needle will not know the glory of Ody's life, his anxious, brave soul, his unfathomable love for the human creature.

I found myself wondering if there was some way of caring for our elderly and ailing pets that didn't simply devolve down to the blue needle. What we need, I thought, are more options for end-of-life care: we need better ways to interpret and manage their pain, more options that might ease them gently toward death or even allow a natural death, clearer ways to think about their quality of life.

ANIMAL HOSPICE

We need something similar to human hospice but for animals. As I began researching, I quickly discovered that I was not the first to have this idea. Animal hospice—often also called pet hospice—already exists. At the beginning of Ody's decline, hospice was under the radar enough that I had missed it, but even by the time I was finishing this book, animal hospice was much more firmly established and more broadly available. In some respects, I did provide hospice care for Ody, but had I known a year ago what I know now, I would have been able to make even better choices and would probably have been able to make the end of Ody's life more comfortable.

The term "hospice" has been used in the context of animal care for at least two decades, and as early as 2001, the American Veterinary Medical Association tried to establish industry standards by issuing its Guidelines for Veterinary Hospice Care. Still, many pet owners and even veterinarians are largely unaware of hospice, and the range of options for pain management and comfort care in our ill and dying animals has remained quite narrow. Now animal hospice seems finally to be gaining a bit more momentum and visibility. About once a month I go online and search for available animal hospice resources, and each time I look the range of services I find being offered has expanded.

Why now? The emergence of a pet hospice movement is certainly, in part, an assimilation of human hospice. It is also a manifestation of changing attitudes toward animals: increased sensitivity to their cognitive and emotional complexity translates into a stronger sense of responsibility toward them and, in particular, more careful attention to the manner of their death and treatment of their pain. Pet demographics have also changed. We now have an increasingly large cohort of elderly, frail, and ill companion animals. As better veterinary treatments allow animals to live longer, this cohort will continue to expand, and greater numbers of animals will live long enough to develop illnesses and disabilities. At the same time, advances in veterinary medicine mean that pet owners will face an ever more complex array of treatment options and, as in human medicine, more and more points at which to decide yes or no. The terrain of end-of-life care for animals is becoming much more complicated.

Although animal hospice has grown, its precise nature remains a bit obscure. At the moment, animal hospice represents a whole range of approaches to end-of-life care, and within these various approaches there are underlying philosophical disagreements about what constitutes a good death for an animal. As the field of hospice expands, more clarity about its specific goals, practices, and philosophical commitments will likely develop. For now, then, "animal hospice" can serve as shorthand for those approaches to animal dying that seek to maximize both quality of life and quality of death.

WHAT EXACTLY DOES ANIMAL HOSPICE INVOLVE?

Unlike other forms of veterinary care, hospice is directed specifically at the care of dying patients, and the goals of treatment are explicitly palliative, not curative. Hospice care aims to provide a terminally ill or dying animal with comfort and palliative care so that the animal can live out its final time with as much quality of life as possible. Like human hospice, animal hospice offers an alternative to aggressive treatments for terminal illnesses such as cancer and kidney disease. It is a mindset adopted by the animal's caretaker and also by their veterinarian. The focus of attention shifts from cure to care, and death is openly accepted as the inevitable outcome. Although hospice care often does prolong the life of an animal, this is not its goal. As Kathryn Marrochino says, quoting Tom Wilson, animal hospice means saying to our animal, "I will be there for you when the time comes, and I will dance with you until the end of the song."

Most practitioners view animal hospice as an alternative to premature euthanasia: sick, disabled, and old animals often have a good deal of quality living to do but are euthanized too early simply because owners and vets aren't aware of other options. But the therapeutic option of euthanasia is always available.

Veterinary practices don't have the range of expertise that human hospice can offer—usually the "care team" is just a veterinarian and a veterinary technician. Still, animal hospice can maintain the emphasis on a team approach—with the team including the veterinarian, the veterinary technician, the (human) client, and the patient (the animal). And veterinarians can be proactive in educating clients about other therapeutic options (e.g., rehabilitation, homeopathy, acupuncture) and avenues for emotional support (pet bereavement counseling, animal chaplaincy), helping people to expand their caregiving network.

Animal hospice largely takes place at home, with occasional visits to the vet if necessary. And an increasing number of veterinarians are offering in-home services, which are usually by far the most comfortable for the animal and also the most practical for the caregivers. The vet will typically come to the home for an initial consultation, which involves assessing the animal's condition and talking with the pet owner about goals for treatment and care. The veterinarian and pet owner can then create a treatment plan. The vet will help the pet owner understand what the animal needs and will teach basic techniques such as administering medicines and even giving injections. Most of the veterinary-based hospice services that I explored will support pet owners who desire a natural death for their animal but also provide euthanasia services.

The key practical aspects of animal hospice care include relieving pain and discomfort (which can involve administration of drugs, massage, physical therapy) and maximizing pleasure for the animal (e.g., through social interaction, companionship, mental stimulation, play, walks, human touch, delicious foods). Care for an ill or aged animal can involve turning, bathing, and assistance with urination or defecation. An animal may need a special diet and even help eating or drinking. Pet owners who commit to hospice care will need to become educated in behavioral signs of pain and distress and may need to learn some basic techniques such as giving injections. The burden of care in animal hospice falls squarely (and heavily) on the animal's Person, and this can shape the landscape of options and choices. And it must be said: hospice care is not for everyone. It can require a substantial commitment of time and money, a certain level of skill and learning, and a willingness to take on a big responsibility. Hospice care done in a half-baked or careless manner can easily result in unnecessary suffering for an animal.

In human medicine, the primary physician typically refers a patient to hospice, and thus to a physician specially trained in palliative care, and to a whole network of caregivers who specialize in hospice work. For animals, there is generally no hospice to which to refer patients, and there is no ready-and-waiting cadre of palliative-care vets. Your regular vet will usually need to shift hats and become your hospice vet. But this is not an easy transition for vets to make or, for that matter, pet owners. Few vets have specialized training in palliative-care techniques, and many are unfamiliar with or uncomfortable with hospice care.

HOSPICE AND PALLIATIVE CARE

Hospice is a type of palliative care aimed specifically at the dying patient. Yet there is no sharp line between hospice and palliative care, and in relation to animals, the two are intertwined. Some hospice advocates even think the two terms should be linked, so that we speak of "hospice and palliative care for animals" as one integrated concept. When we think of hospice and palliative care as a continuum, we can recognize that hospice care may be appropriate even for an animal who has, relatively speaking, a long time still to live and who is not yet actively dying.

Human patients typically enter hospice when they have six months or less to live. Relatively good survival statistics exist for most human diseases, so physicians can make reasonable projections about how long a patient will survive. Far fewer statistics are available for average survival in animals, given particular diagnoses, so it is difficult to judge when exactly an animal is terminal. The time frame of end-of-life care is thus much more open-ended. Furthermore, when scaled to an animal's life span, six months is quite a long time. A comparable "very end-of-life" time frame in dog or cat years would be about a month, but often palliative and hospice care for an animal will be appropriate for a much longer period of time than a month, or even six months for that matter. With animals, hospice care really begins when we judge that cure is either impossible or unduly burdensome, and when we shift our mindset from "fixing" to simply maximizing comfort and quality of life.

One of the most vocal advocates of animal hospice is veterinary oncologist Alice Villalobos. She calls her approach "pawspice" to differentiate it from human hospice, with which she claims philosophical disagreement. Instead of waiting until a patient has only a short time to live, as human hospice does, we should provide hospice to animal patients as soon as they are given a diagnosis, according to Villalobos. And rather than see hospice as simply waiting for death, we should treat hospice as a much more proactive application of palliative care. Animal hospice, for her, primarily provides an alternative to further aggressive treatment, particularly for various cancers. The goal of pawspice is to help animals die without pain and suffering and to help human companions figure out when an animal's quality of life shifts from acceptable to unacceptable. She sees pawspice as filling in that crucial time between diagnosis of terminal illness and death. Although animals with cancer may be the most common candidates, hospice and palliative care are appropriate for all ill, extremely aged, and dying animals over an indefinite period of time.

"It is really important to let pet owners know that their animal will die and that they need to know and deal with it," she explained. Vets aren't adept at talking with clients about animal death. There is a can-do mentality in veterinary medicine, just as there is in human medicine. And when it is clear that they "can't do" any more, vets and clients often turn to euthanasia. There has to be a different way, she says, a change in direction from trying to cure to simply trying to comfort and care.

A CRITICAL NEED FOR HOSPICE

Most veterinary offices don't explicitly offer hospice services. I asked a few veterinarians in Longmont whether they provide hospice care, and they seemed genuinely puzzled. "I'm not really sure what you mean by hospice," one told me. Most vets are comfortable with decisions to forgo treatments with minimal value to the animal and are willing to help their clients try to maximize quality of life for an animal. Still, if pet owners don't know about hospice or palliative care, they will not ask. They may see their only options as pursue treatments or euthanize. If veterinarians don't know about hospice, they won't counsel or offer options, and if they don't have sufficient skill in palliative care, they will be unable to help effectively keep dying animals comfortable. As a result, premature euthanasia will remain a too-common outcome for old or terminally ill animals. As palliative-care vet Robin Downing said to me, "It is horrifying how many cats and dogs are dying without the owners really having an option other than euthanasia."

The need for animal hospice and effective palliative care is vastly greater than the amount of services available. Nonetheless, the number of veterinarians interested in hospice is growing, and we can even see some effort over the past decade to professionalize animal hospice and to establish a concrete sense of agenda and scope of practice (and also, it might be said, a sense of ownership). As I noted earlier, the AVMA has sought to establish industry standards with its Guidelines for Veterinary Hospice Care, aiming above all to ensure adequate pain management and regular monitoring of animals by trained staff.

Two veterinary groups have been particularly active in defining and promoting hospice. The first group, the American Association of Human-Animal Bond Veterinarians, defines their mission as advancing "the role of the veterinary medical community in nurturing positive human-animal interactions in society." The association has taken an interest in hospice, which they define as "a system which provides compassionate comfort care for patients at the end of their lives and also supports their families in the bereavement process." The other group, the International Association for Animal Hospice and Palliative Care, promotes hospice as an alternative both to premature euthanasia and to suffering that can result from isolating an animal in a crate in an animal hospital or from inadequate treatment at home. Hospice is a way to prolong not length of life itself but length of quality life. This association seeks to raise awareness among veterinarians and pet owners about the availability and positive benefits of hospice care.

Animal hospice and palliative care seem to be on the upswing. One veterinarian I interviewed estimated that there are currently between sixty and seventy-five veterinary palliative-care services in the country and the number is growing monthly. Another estimated that the number of animals treated annually by practitioners specializing in some form of end-of-life care is currently somewhere between one and ten thousand in the United States, approximately a tenfold increase from ten years ago.

HOSPICE VOLUNTEERS

Volunteers are a key component of human hospice work. They help provide social support for patient and family, help patients with daily tasks such as cleaning and eating, and offer a respite for family members who need a break from caregiving. In contrast, volunteers are almost nonexistent for animal hospice. This is unfortunate because help—even a simple as having someone come check on an animal and perhaps administer fluids or change bedding while a pet owner is at work—can make hospice a more realistic option for many people.

As one model of what volunteers could offer to animal hospice, we can look at the Pet Hospice Program run by the Argus Institute at Colorado State University, which is currently the only veterinary school–based hospice in the country. Pet Hospice serves Colorado State's veterinary hospital and is mostly staffed by volunteer veterinary students. They provide assistance with decision making and support before, during, and after euthanasia, as well as grief counseling. They will meet with clients and their animals at the veterinary hospital and will be present during veterinary visits, if the client desires. Volunteers also visit homes to help provide hospice care, often administering medications. They don't perform euthanasia but will be present for that process to provide support. Not only do animals and their People benefit from this program but so do the veterinary students, who gain valuable hands-on experience caring for animals and dealing with end-of-life issues such as pain management and bereavement.

Another innovative program that incorporates the work of volunteers is the "Fospice" program run by the San Francisco Society for the Protection against Cruelty to Animals. As far as I can tell, this is the only hospice program specifically for terminally ill shelter animals. Fospice is a foster care program in which trained volunteers offer to take a cat or dog with life-limiting illness into their home and care for them until they die. These life-limiting conditions include renal failure, early heart failure, and non-painful types of cancer such as slow-growing lymphoma. The animals remain the property and primary responsibility of the shelter, and all veterinary treatments, medications, and special foods are provided by the shelter, as are euthanasia and cremation services. The Fospice volunteers provide the daily care, love, and companionship that the animals desperately need. And they are entrusted to judge whether and when euthanasia is appropriate for the animal.

FREE-STANDING HOSPICES

Although almost all hospice currently takes place in homes and veterinary offices, there are a few free-standing residential animal sanctuaries that provide hospice care for elderly, ill, and injured animals. Some of the animals that wind up at sanctuaries have been placed there by loving owners who simply cannot handle the needs of the animal; some of the animals have been abandoned and by one bit of luck or another wind up being rescued rather than killed. What I find most inspiring about these sanctuaries is the recognition that "broken" animals need love and affection as much as any others, and that animals who we might have been labeled hopeless often have a great deal of living still to do. Typically, these sanctuaries are run by ordinary people, not veterinarians. And what I found—though this certainly may not apply to all sanctuaries—is a stronger-than-usual commitment to holistic veterinary care and to natural death as opposed to euthanasia.

BrightHaven Holistic Retreat and Hospice for Animals, founded and run by Gail Pope, is a nonprofit animal sanctuary in Santa Rosa, California for senior and disabled animals. Although BrightHaven does not identify itself as a hospice, per se, it does offer itself as an alternative to euthanasia: many of the BrightHaven animals come straight from under the needle— they have been diagnosed as terminal and hopeless and their humans either cannot or don't want to care for them. The sanctuary offers these animals a chance to live and die in a loving environment. The focus is on a holistic approach to animal care and death. As Gail told me during a phone conversation, "In the traditional veterinary world what happens is that you get a prognosis from the vet that no more care will help and then you wait until it is time for euthanasia. This is a blank time, between the end of care and euthanasia. But with hospice, this period of time becomes one of healing; the animal is cared for differently." She says the animals know that they are dying: the animal who is dying will eventually come to the center of the house and lie down on a bed there. The other animals will gather around. Animals, she says, "are more evolved beings. They live in a different energetic world; they take on death with equanimity."

Another residential hospice is Angel's Gate Hospice for Animals in Delhi, New York, founded by Susan Marino. Angel's Gate takes in various special-needs animals and focuses on rehabilitating them, as well as simply providing a home for them to live out their lives. Angel's Gate, like Bright-Haven, shows a preference for integrative and alternative therapies, raw diets, and natural death. Euthanasia rates at Angel's Gate are reportedly very low—somewhere around 5 percent.

A DARKER KIND OF HOSPICE

As consumer interest in hospice increases, so do the opportunities for exploitation. For example, a few so-called hospices are really just euthanasia services in disguise. If you look carefully, you see that the only service really being provided is a quick death.

Several veterinarians and others close within the animal hospice world talked, darkly, about the existence of for-profit hospices (with the emphasis on for-profit). These sound like small warehouses staffed by veterinary technicians, with an occasional walk-through by a vet, and are designed for a wealthy constituency who can afford to place their ailing or extremely aged pet in a permanent home away from home. Who has time, after all, to constantly check on and clean up after a very sick animal? These places, if they exist, might serve a certain kind of pet owner who feels responsible for their animal and is willing to pay to assuage their feelings of guilt.

I have been unable to verify the existence of such places. Whether real or not, the specter does remind us that hospice, though it sounds loving and compassionate, could easily become something ugly. It is not hard to imagine that as demand for animal hospice grows, some unscrupulous humans will capitalize on it. The market for pet bereavement and death memorabilia is already bustling, and one can too easily imagine the entrepreneurial possibilities. The fact that the AVMA has established guidelines is a good first step, but unfortunately, these only apply—purely voluntarily—to veterinarians who decide to offer hospice services. No laws forbid an enterprising individual from simply opening up a hospice shop.

One of the concerns of veterinary groups is that people who are not vets will decide that opening up a hospice sounds like a great idea: "I love animals and I'm going to open a hospice to help care for the dying." One person active in vet-based hospice told me, "Some hospice workers scare me because they adopt dying animals to just 'be with them' until they die. They aren't veterinarians. They don't know the first thing about managing pain." We might think that veterinarians are simply being proprietary, but I wonder if their concerns have some validity. Knowing how challenging good pain management can be, for example, it seems important to have people with an appropriate level of training attending to the needs of animals. Neither Susan Marino of Angel's Gate nor Gail Pope of Bright-Haven is a veterinarian. And as Gail told me, quite openly, they have been criticized for the way they approach pain management—through homeopathy. Angel's Gate, too, has been the center of some controversy, including an undercover "investigation" by People for the Ethical Treatment of Animals.

I sense a certain amount of animosity and territorialism within the animal hospice world. My take on things is this: the people who actively work in animal hospice—veterinarians and nonveterinarians alike—do so because they are passionate about the welfare of animals. The intentions are (almost) all good. The problem boils down to philosophical differences about pain management and natural death, as well as broader philosophical disagreements about the role of companion animals in our lives. To oversimplify grossly and to caricature: veterinary-based hospice tends to be conventional, scientific. Pain medicines are titrated with a calculator, and signs of pain are based on scientific models (in such form as these exist). The end point, for vet-based hospice, is generally though not always euthanasia. Sanctuary-based hospice, conversely, leans toward the more holistic and spiritual, with animal communicators helping to decipher the desires of dying animals and with aromatherapy and homeopathy used as pain control. Sanctuary-based hospices often eschew euthanasia and even conventional medical treatments, including pain medication. The end point is, ideally, "natural" death.

NATURAL DEATH

Animal hospice is not an alternative to euthanasia, and I want to say very clearly that I believe euthanasia very often becomes, at some point, the compassionate course of action. Still, hospice is often embraced by those who desire a natural death for their animal, and even those who favor frequent use of the blue needle should take time to consider what "natural death" means and whether it might be of some value to our animals.

My sociologist friend Leslie Irvine called me recently because her twenty-year-old cat Ms Kitten was dying, and she wondered whether I would like to visit with Ms Kitten before she passed. Unfortunately, Ms Kitten died before I got there, but Leslie talked with me about Ms Kitten's death and shared an account of Ms Kitten's last few days.

> When she began to show signs of declining, Marc and I agreed that we would not euthanize her unless she was suffering. We wanted her to exit life naturally. We did not need her to hurry. We were not afraid of seeing her die. We talked about how we might know the point of suffering; cats can be very stoic. We watched her closely for signs of pain. We also watched to see if she wanted to withdraw. To the end, she seemed comfortable being near

us. She never flinched or had trouble breathing. I wonder now if the way she wandered around on Thursday was a form of "terminal restlessness." But it did not seem like distress or struggle. It was more like what she had always done, only more slowly and with pauses here and there.

Leslie wanted Ms Kitten "to exit life naturally." When I asked her why, she said that she believed we should let an animal live out its full span of life. And I find something dignified and right about Ms Kitten's death. She died as close to a "good death" as I can imagine: she did not seem to be in pain, she was surrounded by loving companions, and she just stopped breathing. For Leslie and her husband, a natural death was preferable to euthanasia, though they were willing and ready to take that route had she been clearly suffering.

Dying can be a time of opportunity. People change as they are dying; human development continues up through the very end of life. And the dying process can be a time of profound personal growth (and family growth, too). For some people, the value of natural death is explicitly spiritual, for instance, as a time of transition from one kind of being to another. Although we cannot know what goes on inside the minds and hearts of our animals, perhaps we should be open to the possibility that they, too, might experience something profound as they die.

This is the appeal of natural death—and this is very much what I would wish, all else being equal, for Ody. I appreciate the attention to the spiritual dimensions of animal death, and the sense that ritualizing the passing of our animals is a way to affirm the value of their lives and the strength of the human-animal bond. But I also think "natural death" can be taken too far, if we use it to denote some particular kind of death, as opposed to "unnatural death" (or some such). All death is natural, whether it takes place in the wild under the stars, on a cold metal table, or on an oatmeal-colored dog bed under the piano. Death by euthanasia is every bit as natural as death by starvation, dehydration, or multiple organ failure. And we have, from the moment they enter under our care, such utter control over every aspect of our animals' lives and deaths that I'm not sure we can ever really, truly "let nature takes its course."

We can certainly simplify the moral terrain by placing natural death and euthanasia in opposition. If we commit to natural death in this sense, then our responsibilities are limited to standing by our animal and offering comfort as they die. But moral simplification is not what we want. The same kind of caution applies, more broadly, to hospice care. When hospice and euthanasia are presented as stark and opposing choices, we've oversimplified the moral terrain, at the expense of our animals. Consider this cautionary note from Johnny Hoskins, author of *Geriatrics and Gerontology of the Dog and Cat.* Veterinary staff, he says, should inform owners that hospice care ("terminal life care") is available and exists as an alternative to immediate euthanasia. But presenting hospice care as an alternative to immediate euthanasia may not always be in the animal's best interest. "In virtually all cases of progressive incurable illness, the animal reaches a point at which suffering is simply not necessary or fair. . . . If hospice care is perceived solely as an alternative to suffering then certain owners, eager to find any reason to avoid euthanasia, will happily grasp this as a guilt-free means to reject the euthanasia option." Hospice is not, to reiterate, an alternative to suffering. The goal of hospice is certainly to minimize suffering wherever possible, but sometimes suffering outfoxes all our best efforts.

Robin Downing told me that she has sensed, within the animal hospice world, an undercurrent that espouses translating and applying the exact same principles of human hospice to our animals, including a prohibition against euthanasia. "Unconscionable!" she said to me. "Euthanasia is available and we have the liberty of applying it with animals. And we must. It is the ethical thing to do." In her view, the pendulum has swung from "let's euthanize today" to "natural death." And she doesn't like it. We need to move back toward the middle, toward a more balanced place. I asked her how many animals she has had die natural deaths. In twenty-five years of practice, she said, "not many." And she has been very sensitive to the idea of allowing animals to live until it is really time for them to go to heaven. It just doesn't happen very often that natural death is in the interests of the animal.

I asked one of our local vets how many of his clients express a desire for a natural death for their animal. "Surprisingly, not as many as you think," he told me. His feeling is that natural death is very rarely a good death and that "natural causes" are usually pretty unsavory for the animal. Kidney disease, for instance, is the second leading cause of death for cats and is not a fun way to go, what with the possibility of dehydration and seizures from toxin buildup.

It is dangerous to assume that natural death is always better for an animal. But is it often preferable? We sometimes hasten death unnecessarily. If, as death unfolds, the animal is suffering considerably, then hastening death is perhaps the humane path. How we define and judge "considerable suffering" is, of course, the nub of an irritating problem, and some people may be better at reading signs of animal suffering than others, or some animals better at signaling pain. Our own subjective values as pet companions will undoubtedly color our vision; if we are strongly committed to natural passing, perhaps we will "read" signs of suffering differently—as "signs of a transitioning soul" rather than "signs of an animal in torment."

A NOTE ON LANGUAGE

What is the most fitting language to use in describing a death that occurs without euthanasia? "Natural death" is the most obvious choice and has intuitive appeal. "Natural death" carries positive connotations of avoiding unnecessary hastening of death, avoiding the violence of the blue needle, letting nature take its course. But there are problems with this terminology, too, and many working with animal end-of-life care are uncomfortable with it.

One danger of the term "natural death" is that it suggests that death should be unassisted—yet a good death will very often require our intimate and persistent help, with various interventions to manage pain and keep an animal comfortable. A passive approach to death, where we just sit back and watch and don't interfere, may be natural but is rarely good for the animal. According to some advocates, it is important to ensure that animal hospice does not become associated with a denial of treatment, as happened early on in the sphere of human hospice. This has been a very hard stereotype for human hospice to shed and has slowed the growth of the human hospice movement. The Nikki Hospice Foundation for Pets, the first nonprofit to promote the idea of home-based hospice, favors the consistent use of "hospice-assisted natural death." This, I think, is pretty good.

We still have the difficult question of what language to use to describe death that we have deliberately brought about with a lethal injection. Do we call it euthanasia? Terminal palliation? Palliative sedation? We'll return to this issue in the next chapter.

SUFFERING AND QUALITY OF LIFE IN ANIMALS

Quality of life (QOL) is of enormous interest in bioethics and human medicine and has been used as a tool for evaluating an individual's overall well-being, as judged from their own perspective. The question of QOL is often the basis for decision making about treatment, palliation, and death. Beginning in 2000, QOL entered the vocabulary of veterinary medicine, and it has the potential to be an important tool as we navigate, with our animals, end-of-life care.

Veterinarian Frank McMillan has been particularly active in championing the use of QOL in animals. Quality of life, he says, is closely related to welfare and psychological well-being, except whereas welfare and well-being are external evaluations—humans judging the quality of an animal's life—QOL is meant to be "a view from within." The difficulty with animal QOL is immediately apparent: we are the ones, ultimately, who make the QOL assessment, and to do so we must try to get inside the head of the animal and figure out just how this view from within looks.

In human medicine, QOL is supposed to represent a subjective assessment of well-being and happiness—how do you, the patient, judge the quality of your physical and emotional life?—but in practice, the concept is most often used in relation to those who cannot articulate such judgments for themselves: neonates, infants, the mentally disabled, the senile, the comatose, and the severely ill. We instead rely on proxy QOL assessments, which is essentially what we have in dealing with animals. And proxy assessments are vexing. McMillan notes that within human medicine the accuracy of proxy ratings has proven uneven. In studies of adolescent patients, for example, proxy assessments of emotions and subjective states made by their parents did not map particularly well onto self-assessments made by the adolescents. We may know our dogs and cats better than our teenagers, but it must still be said that cross-species proxy judgments involve substantial uncertainty. McMillan believes that behavioral research techniques—preference testing, aversion learning, and demand curve analysis—may eventually allow a better understanding of the private feelings of animals, and more confidence in the accuracy of our proxy assessments. But for now, we're flying by the seat of our pants.

And what choice do we have? How do we evaluate how our animal is feeling about his or her life? Each animal is unique and has its own preferences and subjective states. Quality of life is highly individualized; there is no "normal" or average, in terms of what makes animals happy. McMillan proposes what he calls an affect-balance model. Taking the major factors contributing to QOL for animals—mental stimulation, health, unpleasant emotional experiences, and degree of control—we can basically assess whether positive affect (pleasure) outweighs negative affect. Quality of life represents a continuum, from very good to poor.

McMillan suggests making a list of factors in your pet's life that affect QOL. What matters to an animal will be based on a pleasant or unpleasant affect associated with each feeling and experience. Pleasant affect includes both positive emotions (joy, sexual stimulation, mental stimulation) and positive physical sensations (good food, social closeness). And negative affect includes physical sensations such as pain and hunger, as well as negative emotions such as anxiety, loneliness, and depression. Unpleasant affect, he says, has a stronger grab on animals than pleasant affect—it "weighs" more on the QOL scales. The most urgently threatening stimuli are associated with the most strongly unpleasant affects (e.g., the extremely unpleasant sensation of being deprived of oxygen; the extreme pain of tissue damage). So, basically we take all the things that we think influence QOL for our animal, both good and bad, and assess the relative intensities of these affects and the amount of time they last. We then try to determine the relative balance of positive and negative. When the balance shifts strongly to the negative, we then think about whether continued life is of value to the animal.

What to put on Ody's QOL scales? His pleasures include lying on his bed (sleeping or just watching); taking short walks; marking on the neighbor's roses; eating cheese and his special homemade food of hamburger, rice, and carrots; licks on the muzzle from Maya; attention from us (though he seems ambivalent about this, honestly); saying hi to people, especially children. He seems to enjoy going out into the backyard and just taking stock of things, standing with his nose up to the sky. His list of displeasures is quite long: he seems to be in some pain in his back legs, and his teeth just have to hurt (though the vet thinks not so much); he seems distressed when his body won't do what it should; he lives in fear of being attacked by Topaz; he might be lonely when the rest of the human-canine family trundles off for adventures (I really can't tell); his deafness is pretty isolating; his poor eyesight seems to make him nervous; he doesn't seem to enjoy food nearly as much as he used to (he will only eat his hamburger meal; even hotdogs are losing their appeal). I have no idea how to weigh these "affects," though I feel pretty confident that Ody's scales still tip toward pleasure. Will I be able to tell when the balance shifts?

It is easy to say, as an able-bodied person, that you would not want to live as a quadriplegic or some such, but able-bodied people who become disabled often report very high quality of life. And plenty of people born with disabilities find ways to live fulfilling lives—often finding more happiness and satisfaction than people without obvious disabilities. In a similar vein, we should be careful jumping to conclusions about what conditions animals would or wouldn't want to live with. For example, people might think that a cat would rather not live at all if it had to be blind. But read *Homer's Odyssey* by Gwen Cooper and you will get to know a blind cat so full of spirit, excitement, and love that his disability almost seems a gift. And read *Amazing Gracie* about a white Great Dane who has congenital deafness but lives a full and joyful life. Plenty of dogs lose a leg, or even two legs, and adapt extremely well to their disability. Surely there are injuries and disabilities that would make life unbearable for an animal, but we should be circumspect in making such judgments.

At the same time, many of us have had this experience or heard about it: a friend or acquaintance who holds on to their pet too long, who ignores blatant clues that the animal's QOL is poor, where you look at the animal and just know that they would rather not be there. Love can sometimes blind us to reality. And this makes me think that although I am in the best position to make QOL judgments about Ody, I need to seek input from more "objective" observers. I fear that I might minimize Ody's problems, simply because it is too painful to contemplate his death. When my friend Liz comes over—and she knows Ody better than any of my other friends and has known him for most of his life—she sounds concerned. "Oh God," she'll say. "It's so sad. Ody is just a husk of his former self." And Chris says on some days, as if to desensitize me through repeated exposure, "I think Ody is getting close to his time. Don't you?"

THE PAWSPICE SCALE

Alice Villalobos's Pawspice "HHHHHMM Q of L Scale" is a shorthand method for assessing how an animal is doing and is less cumbersome than McMillan's QOL affect-balance scale. What I appreciate about the Pawspice scale is that it offers concrete considerations for evaluating one's pet (hurt, hunger, hydration, hygiene, happiness, mobility, and more good days than bad), based on a 1–10 scale (with 1 being low, 10 high), and even a specific score by which to make euthanasia decisions: 35. A score of 35 or more indicates an acceptable QOL; an unacceptable score (<35) suggests that euthanasia may be the best option. This shifts attention from the animal's own subjective assessment of QOL to our evaluation of his physical well-being, and removes some of the guesswork. But of course the aura of concreteness belies the painful subjectivity of our judgments. Assessing QOL, as we've seen, is really a gestalt.

Using the Pawspice scale, I would judge Ody today, on November 26, 2010, as follows.

Hurt: 6. I don't believe that Ody is in much pain, but I don't feel confident about this judgment. The vet tells me that Ody's problems with his legs are neurological and that he actually has very little feeling back there. (This is probably why he poops on the floor—he isn't really aware that he is having a bowel movement.) But sometimes when I try to gently massage Ody's body he jerks up and tries to bite or simply struggles up and stalks off. This, to me, suggests some discomfort. Also, he suffers some emotional pain, I think, because he often seems quite anxious.

Hunger: 6. Ody is becoming very thin and often refuses his dog food, but he has a very good appetite for treats and his homemade hamburger-rice meal.

Hydration: 9. Ody still drinks readily. I put 9 instead of 10 because I don't think he gets enough water (because of his lack of mobility).

Hygiene: 4. Accidents in the house are becoming more frequent. Ody sometimes poops and then either lies in it or steps in it—and this is becoming more and more frequent. Last night he defecated and urinated both, which is a first. And he didn't bark to go out. Also, his bladder seems to leak a bit because when he sleeps on the couch there will often be a wet spot underneath him. His breath is horrible, but I don't think this bothers him in the least. His hair is short, so matting isn't a problem, and though he has some dandruff and sheds a lot, his coat still looks reasonably good.

Happiness: 6. How do I judge this? He seems anxious and lonely a lot; he will come into a room and just stand, as if he wants something. He seems interested in what the other dogs are doing—when they bark at passersby, or wag tails because they know a treat is on the way. And he seems to enjoy his small walks up to the neighbor's house.

Mobility: 4. He can still climb on the low couches, can still go for short walks, and can move around to eat and go outside. But it is not easy for him to stand up from a prone position and he sometimes topples over. And he cannot stand erect for more than a few seconds before his hind end sinks toward the ground.

More good days than bad: 7. He doesn't really have good and bad days— most days seems pretty much the same (though there is a gradual decline, noticeable over a span of weeks). There are days when we have major poop disasters in the house, but are these bad for him, or just me?

His score: 42 (out of a possible 70). This is actually lower than I expected and makes my stomach ache a bit. He's inching down toward 35.

ADVANCE DIRECTIVES FOR PETS

Research suggests that a large majority of people would not want their lives prolonged if their prognosis was very poor or if it meant being permanently and seriously incapacitated. Yet once within the medical system, the default option for people who are too sick to speak for themselves is to err on the side of more treatment rather than less, no matter how bad the outlook. The advance directive has become an important element in efforts to improve end-of-life care and death, particularly in helping people avoid a vicious spiral of unwanted care and prolongation of life. An advance directive (also called a living will) is a written document prepared by an individual to tell others what kinds of medical decisions they would like made, in the event they become unable to speak for themselves. The directive might specify, for example, that an individual refuses cardiopulmonary resuscitation, should they enter cardiac arrest. Or it might specify that they would not want to be kept alive in a persistent vegetative state. Advance directives emerged, more than anything, as a way for patients to say no to invasive and futile treatments. Yet they can also specify that a patient wants every intervention at hand.

Advance directives have been controversial because instructions such as "no heroic measures" can be dangerously ambiguous. And studies have shown, over and over, that doctors and families routinely ignore directives. Nevertheless, the use of advance directives is slowly growing. The single most important function of advance directives, according to advocates, is that they get people thinking and talking about death and the dying process. Very few dying people actually talk with their doctors or families about dying, and by encouraging frank discussion, advance directives have been shown to improve patient welfare dramatically and to decrease suffering for patient and family alike.

Hoskins suggests, in *Geriatrics and Gerontology of the Dog and Cat*, the use of a modified advance directive for animals. As with human advance directives, the key may lie, more than anything, in encouraging people to think and talk about their animal's death. Obviously we can't talk with our pets about their preferences and hopes and values at the end of life, but filling out an advance directive will encourage us to consider what's important to us and our animals. Hoskins recommends including the following kinds of questions: Who is the primary caregiver? Who else is involved in care and decision making? Do you want the final hours to be spent at home or at a vet hospital? What activities are most important to your pet? Then you might have a list of factors that speak to quality of life and rank them based on how strongly they might cause you to consider euthanasia for your animal. For example, you might rank reduced mobility, social withdrawal, reduced appetite or refusal of food or water, need for daily medications, need for nursing care, loss of vision or hearing, incontinence, and chronic pain. This is similar, in many respects, to our QOL scales.

The benefits of advance directives for our pets are many. The directive (as the name suggests) begins and nurtures the difficult process of thinking through, for yourself and your animal, what a good death means and it reinforces the reality of death. It can make it easier to be level-headed about decisions, and it promotes conversation among family members and between pet owners and veterinarians. We may find that our vet doesn't share our values or preferences about animal death and we may also discover surprising differences among family members. These conversations can also allow children time to understand and process their feelings about the death of a family pet. And the advance directive can give caregivers some sense of what might be involved in dealing with a dying animal.

BEING A CAREGIVER

Some of McMillan's QOL factors, as you can see, are about the animal's quality of life, and some are more about the owner's quality of life or caregiving resources (time, money, emotional energy). We could argue that the list should only include items that speak directly to the animal's quality of life. But this is unrealistic, and also unfair. Anyone who has gotten to the point—and taken the trouble—to fill out an advance directive for their pet is strongly committed to their loving care, and the fact is that caregiving resources don't flow from some magically regenerating source.

A hospice flyer from Colorado State University's Argus Institute lists, after "Consider your pet's quality of life," "Consider your quality of life." They say to ask yourself the following questions:

How much of my time will go toward taking care of my pet?
How much will it cost to take care of my pet?
What other responsibilities do I have in my life (job, parenting)?
Who else (partner, children, or pets) do I need to consider?
Which family members or friends can help?
What other stresses do I have in my life right now?

According to studies of human hospice and end-of-life care, the average caregiver provides four and a half years of care, and three out of four caregivers are women. I'm not aware of comparable studies of human caregivers for dying animals, but it is safe to say that the period of caregiving—particularly if the approach to death is hospice oriented—will sometimes be significant and the demands intense.

So, how do we balance the needs of our animal with the many other pulls on our time and energy? There is no formula for just how much care is the right amount and how much sacrifice is necessary. I have heard plenty of touching stories of people who quit their jobs or take long leaves of absence in order to be present fully for their dying animal. At the same time as feeling touched, I also feel guilty and resentful. I can't live up to this standard. I can't put childcare aside; I don't want to quit working. My suspicion is that a great many of us feel guilt, no matter what choices we make. We have never done as much as we could have.

One of the most important services offered by a human hospice team is respite care—where a hospice worker comes and allows the primary caregiver a few hours or a day off from caregiving. Unfortunately, there are no respite workers in animal hospice. Friends and family may not understand what is involved in caring for a dying animal and may not think to offer to help. I have often felt a need for a respite from Ody. Stumbling bleary-eyed into Vic's coffee shop in the morning—I have become one of their best customers over the months of Ody's decline—I fantasize about an unbroken night of sleep. But in the larger scale of things, having to get up two or three times a night to let Ody out is tolerable. Having to clean poop and pee off the floors? This I can also tolerate, though I have to admit that it feels harder and harder to keep up with the mess as he becomes less and less able to control his body. But I can adapt. Not being able to leave home for any length of time, since Ody needs to be let out relatively frequently, is also fine. I work at home and most of my family's activities are close by. Still, I wonder how difficult these same tasks might feel to someone who lived alone and had to work full time away from home, or had to travel, or had such a tight budget that replacing some ruined carpet might not be so easy.

APPROPRIATE USE OF RESOURCES

According to a survey by the American Animal Hospital Association, 63 percent of pet owners say "I love you" to their pets every day. This is probably more than most spouses get, and maybe more than some children. A large number of these pet owners even share a bed with their animals—and not just dogs and cats, either. People sleep with pigs, ferrets, rabbits, birds, and all manner of creepy crawly things. It is perhaps no surprise, then, that many people will pursue veterinary treatments for their animals in much the same way they would pursue such treatment for a spouse or child. Anything that might offer our animals a chance to keep living might be worth a try—our dogs and cats are placed on kidney dialysis, undergo chemotherapy, have hip replacements, and are even given stem cell transplants.

If we place end-of-life care for animals within a broader context, does our perspective change? If we look, for instance, at health care as a whole and the need to make health care more broadly available and more sustainable, should we be spending so much money on exotic and expensive treatments for our pets, like unproven stem-cell therapies and expensive cancer treatments that merely prolong life for a few months or a year?

One of my favorite pet-related blogs is Patty Kuhly's *Fully Vetted* (previously called the *Dolittler* blog). One day she featured a guest blog by veterinarian Nancy Kay, whose book *Speaking for Spot* helps pet owners navigate the complex world of health care decision making for their pets. In the book, Dr. Kay arms her readers with knowledge and tools that will allow them to be the best advocates they can be for their pets' well-being, and she focuses a good deal of attention on how to access the most advanced veterinary care. In her guest blog, Dr. Kay describes some of her "fan mail," such as this email:

> I'm annoyed at how dogs have become soooo important over the past 10 years or so. They're just pets. Just animals. . . . I recently read a book about an African village, and the hard life they have, and the poverty. I found it so shameful that they live like that, while America's dogs are often dressed in designer clothes, waited on hand and foot, given the best medical care, the best food, cooed over, etc. . . . What the hell has happened to America?

If it were a simple equation—a thousand dollars to perform chemotherapy on my cancer-ridden dog or a thousand dollars to save starving African villagers—then perhaps the answer would be straightforward. But this is not how the equation falls out. I'm reminded of the childhood admonition: "Eat your peas; there are children starving in Africa." To which a snide child might reply, "Then let's put these peas in an envelope and send them to those poor children." But unfortunately, US peas don't easily translate into African aid. (The moral point remains intact, however: there is something shameful about wasting food when many have so little.) In fact, people who have animals as pets choose to spend money on their animals instead of other goods. Perhaps they drive a less expensive car or own fewer television sets—who knows? Overconsumption is a problem for most Americans, not just pet owners. Americans waste an offensive amount of food—but pet owners are no more wasteful than others and maybe less so since pets may make good use of scrap food.

There are certainly better and worse ways to spend money on our pets. Designer clothes and diamond-studded collars for dogs are about as easy to defend—morally speaking—as designer clothes and diamond-studded cell phones for people (and why discriminate against dogs, if style is really at issue?). But though designer clothes may be excessive, good food and good veterinary care are not. It seems utterly appropriate to buy for our pets the "best" food, where by "best" we mean the nutritious food made from high-quality ingredients. "Best" food often is the most expensive— this is true for human food, too. The junk dog food is cheapest, sure. But, if our dogs become obese, diabetic, and cancer-ridden because of how we feed them, isn't it then our responsibility to relieve them of some of the symptoms and sufferings of these ailments? And wouldn't it be better all around—and cheaper in the long run—to feed them the best food in the first place? Preventive medicine is on the whole much more cost effective than rescue medicine.

I have been troubled by the problem of consumption as it relates to my pets. I am especially bothered by the fact that caring for a pack of carnivores requires that I buy meat. And I do feel guilty and excessive when my dogs eat better than most people on the planet. But the solution here is not to stop feeding my dogs well but, rather, to put more energy into good works. I am more bothered by the costs of incidentals, like extra treats and the Frisbees that Topaz goes through at lightning speed but cannot live without (once caught, a Frisbee must be torn to shreds). We may tell our animals every day that we love them, and we may pronounce to friends and acquaintances "I am an animal lover!" Yet there is inconsistency in our feelings toward animals: we spend upwards of $47 billion a year on our pets—for food, medical care, treats, and toys. Yet even as we love our individual pets, some 6 million strays are killed each year for want of someone to love them.

There are also more particular questions about spending money on health and on health care. According to the Humane Society of the United States, dog owners spend, on average, about $248 per year on veterinary visits, and cat owners about $219. Surely some pet owners with aging or very ill pets spend considerably more, and this figure doesn't include costs of medications, but as an average, this figure is fairly small. To put things in perspective, according to the Kaiser Family Foundation, average health care spending per capita for humans in the United States is about $5,711. And, for yet more perspective, the average US consumer spends $457 a year on alcoholic beverages and $2,698 on entertainment. What the hell has happened to America, indeed.

We spend money on what we value most: on what makes us happy and improves our quality of life. And animals make some of us very happy. As Dr. Kay notes, and as a good deal of research supports, the human-animal bond has a positive impact on human beings. Having pets lowers stress, contributes to good health, and generally makes owners happier and more balanced and, thus, more capable of making a positive contribution to human society. Could we say the same about a six-pack of Bud Light and a flat-screen TV?

HOW MUCH IS TOO MUCH?

The most perplexing questions, I think, are those that each pet owner must face: How much should I spend on my animal, and on what kind of care? If our resources are limited—and for most of us they are—how do we balance the needs of our pets against other needs and desires? If an expensive treatment for our dog must come straight out of our child's college fund, then there is some hard thinking to do. No formula exists that makes this calculus easy, and although some people's choices make me cringe—like the individual who refuses to treat his dog's severe and painful dental disease simply because "it's just a dog"—it seems best not to judge too quickly because most people do the best they can and often the reasons for a particular choice are better than we might assume. I have made choices about Ody's care: I have declined diagnostic tests that might tell me more clearly what is wrong with him. And although I can justify the decision based on Ody's welfare, part of the reasoning is financial. It has to be.

It may be that going the route of hospice is generally less expensive than aggressive treatment, and often a need or desire to limit spending will drive the decision to forgo attempts to cure our pets. But animal hospice care is not free and is almost certainly more expensive than immediate euthanasia. So it could be that money will be a barrier to the widespread use of hospice for animals. (Ironically, one of the key principles of human hospice is that money should never be a barrier; hospice care is available regardless of ability to pay.)

Unless you drop your animal off at a sanctuary like BrightHaven— which relies entirely on donations—or happen to live within a few miles of Colorado State's Hospice Care center, you can expect to pay for hospice care, and the more you do for your animal, the more it may cost. Pain medications themselves are relatively inexpensive, and human comfort and love are free, so your basic hospice plan is probably quite affordable. But there are always upgrades. For example, you might have to pay a veterinarian or technician to come to your home to administer drugs or to teach you how to do it yourself. You may decide to pay for animal acupuncture, homeopathy, massage, or physical therapy. Maybe a special diet will be recommended (and these are typically quite pricey). As in human medicine, there will be a tiered system, with the wealthier patients having access to better services. Relatively few people have pet insurance, so the questions of money are even more pointed than they are in human medical care.

Ody's expenses, for one month, when he started to really go downhill (when I would say we shifted into hospice mode): a doggie waterbed ($70); a full blood panel ($160); two in-home vet visits ($80 each); thyrosin ($15); prednisolone ($15); an additional thyroid function test ($100). Of course, the longer I keep Ody alive, the longer I pay for his special homemade hamburger-and-rice, hotdogs, paper towels, and Nature's Miracle stain and odor remover. The vet highly recommends rehabilitation therapy, which is a hundred bucks each visit and laser therapy (which I haven't priced yet), and she thinks I should definitely try acupuncture to see if it helps Ody's mobility (haven't priced this yet, but my guess is another $100, at least, to have someone come to my house since Ody really isn't up to much car travel).

NEW APPROACHES TO ANIMAL DEATH

Vast numbers of animals who would be excellent candidates for hospice wind up being euthanized while they still have a good deal of living to do, simply because their owners do not know that there is another way for animals to grow old and die. Pet euthanasia has become so deeply ingrained in this country that people don't even consider other possibilities. As one advocate of animal hospice told me, the standard veterinary approach is "treat or kill." When an animal can no longer be effectively cured, or when their human caretakers no longer want to pursue treatments, you must then check the box next to "euthanize." You don't (appear to) have any other choices, so your next step in caring for the animal is to throw in the towel. In this sense, hospice is a revolutionary way of thinking about animal death. Animal hospice broadens the possibilities and stimulates creative thinking about how our animals can die and what compassionate care can look like.

I am obviously a fan of animal hospice, and although I strongly believe that hospice is an important alternative to premature euthanasia, I do not believe that hospice care replaces euthanasia or obviates the need for it. At its best, hospice care will fill a gap that premature euthanasia too often fills, will expand the range of options for end-of-life care, and will draw much-needed attention to the importance of palliative care. Together, palliative care, hospice, and euthanasia offer us a Goldilocks approach to animal death: not too soon, not too late, but just right.

Hospice can do many things: it can help expand and fill in that space between diagnosis and euthanasia; it can challenge the dominant narrative of euthanasia by presenting other possibilities for a good death; it can improve quality of life in a dying animal; it can suggest how to prolong life, without simply prolonging the dying process (though this is, admittedly, a crude distinction); it can make the dying process less painful, less protracted, less lonely, and more dignified; it can get people thinking creatively about end-of-life care for ill and aged animals; it can help people deal forthrightly with their grief and denial over the impending death of a companion, so that decisions about treatment can be based on the needs of the animal, not our own needs. Above all, hospice offers us a gentler way down into the Valley of Death, a slow path down that we can travel hand-in-paw with our animal rather than shoving them brusquely off a precipice. In its best iterations, hospice offers us the possibility of a good death, a fulfillment of the sixth freedom, for our animal companions.

The Ody Journal

OCTOBER 25, 2010

Ody had a hard time standing this morning—his right rear leg wouldn't hold up and his rear legs kept crossing over each other. What's especially hard with Ody is that he doesn't seem to be in pain—or at least not in physical pain. Without pain, it is harder to assess his quality of life. He seems to suffer from psychological pain—he's anxious and panting a lot of the time. But Ody has always been anxious.

OCTOBER 30, 2010

He's having a hard time going poop. Watched him outside squatting and stumbling around the yard, his tail quivering. Came in with diarrhea dripping down his rear end. He tried to bite me when I cleaned him off.

He refuses all dog food now, even soft canned food. He will eat: cut up hotdogs, lunch meat in small pieces, and canned doggie "stew." He can't chew any big pieces. His teeth don't work well enough. If I give him too big a piece, he drops it and then has a hard time reaching down to it.

HALLOWEEN NIGHT, 2010

Woke up about 6 a.m. to odd sounds of scraping and sliding. Ody was on the floor in the office, wedged up against the wall, trapped in a corner where there is no carpet. He was trying, unsuccessfully, to get up. I lifted him and he shuffled awkwardly outside. I tried to go back to sleep for a while. At 8 I got up and the minute I walked into the hall I smelled poop. Ody had had an accident on the floor and then stepped in his own feces and tracked little pieces all over the house. I could follow his path through the kitchen, to the place under the piano, to the office, and back. It will be a long day of cleaning.

We cleaned the floors for about three hours, taking up all the carpets we had laid for Ody and washing them; picking up chunks of poop here and there; sweeping and mopping throughout the house. And still, I have this icky feeling that maybe we missed a spot, that there is poop lurking in some corner or another.

Because two of his paws had poop smeared on them, I had to give him a bath. But once I got him, shaking and quivering and panting, into the tub and began filling it with warm water, his feet began to slip and slide. I couldn't lift either of his dirty paws to wash them because he couldn't stand on three legs. And I couldn't hold his body and lift a foot at the same time—he's too heavy. So, there I was, feeling at a loss.

I yelled for help and eventually Chris heard me and came in. He had to perch on the edges of the tub and hold Ody's body from above so that I could wash. Chris thought I was finished when I reached for a towel, so he let Ody go. Ody slipped down and then over onto his back, covering his whole body with dirty poo water and spiking his anxiety level up a couple more notches. Along with the poop I washed away large clumps of red hair. Either Ody is shedding like crazy or he is starting to go bald or he's losing his hair because he's sick. Every time I ran my hand along his wet fur I came away with patches of hair. Ody seeds washing down the drain.

Because Ody's carpets were all in the wash, he had to contend with the hardwood floors today. He's fallen at least three times and been unable to stand up without my help. I feel, today, that the end is drawing closer. Every time I look at Ody it is with concern and pity. I must say "poor Ody," either to him or to myself, at least a hundred times a day. And guests to the house are starting to say it, too. I feel as though I need to heed these "objective" voices because I believe that as primary caregiver I may be too close, I may not see what others see (because I don't want to). I want to see an Ody who can keep on living, who can stay with me.

Chris gently asks, "Don't you think it is getting close?" The "it" . . . we circle around, dodging. Yes, today I have to say that Ody probably had more distress than pleasure. He really doesn't seem to enjoy food that much, hotdogs notwithstanding. And he seems so remote when I try to give him love. I sit next to him many times in a day and just tell him that he's a good boy and that I love him. But he never wags his tail at me as he used to.

Ody seems confused, too. He wanders the house, like a ghost. He'll stop, now, in front of a wall and seem befuddled about how to get turned around. He goes outside and comes back in, over and over, as if searching for something that he cannot quite remember.

Chris and I talked about "how do you know when it's time?"

We thought:

when an animal can no longer get up and has to lie in its own excrement;
when an animal is in a lot of pain;
when the balance of time is spent in a state of distress rather than pleasure;
when there are no moments of joy.

NOVEMBER 1, 2010

His left eye looks funny today. This morning it looked a little infected; now it looks sunken and half shut.

NOVEMBER 2, 2010

Morpheus is the God of sleep, of dreams. I'm thinking of morphine and terminal sedation . . . would that we could simply ease Ody into sleep.

From *Final Crossing*, p. 13: In 1907, anthropologist Arnold van Gennup coined the term "rite of passage." There is a tripartite sequence common to all life's great transitions:

severance: dying to the old way of being;
threshold: the time between worlds;
incorporation: rebirth into a new life.

Joseph Campbell says, in *Hero with a Thousand Faces*: each person's life story (and each cultural myth) is a version of the hero's journey. Each journey comprises a series of great crossings. Stories of heroes are told around a fire. We should tell Ody's story around a fire, too.

NOVEMBER 3, 2010

Ody had ups and downs today. Ups: he actually ate (lunch meat, hotdogs, canned food). Downs: his rear leg situation got worse. For thirty minutes, his rear left leg went stiff, as if he had a muscle cramp. He almost couldn't walk and kept dipping down on his remaining rear leg, which is too weak to support his body. He seemed stressed out by this—he tried to stay close to me, looked really nervous, and was breathing hard. I don't think he was in pain, but I'm not sure. After a while, he leg seemed to relax.

Last night was hard. Ody woke me at least three times; Topaz woke me once (to be let out of Sage's room); Maya once, to go out. Maya went out once with Ody, too, and started barking like mad, right under the neighbor's window. Sage also woke me once, crawling into our bed, and then about a hundred more times after that by poking an elbow in my ribs, suffocating me by throwing her body over mine, and hogging the blankets. I dream of sleep . . . Morpheus help me.

NOVEMBER 4, 2010

Ody actually wanted to go for a walk with us this morning, and although he was slower than ever, he made it all the way to Todd and Deby's, three houses up, and seemed to enjoy the sniffing and the fresh air. Still, I notice that his walk has become very lopsided. He walks crooked and crablike. His hind end doesn't follow straight after the front, but sways out to the side. And with each step, especially of his left hind leg, he dips almost to the ground. His walk is noticeably different today, as compared to yesterday. It worries me.

He also seems very restless and has wandered around the house all morning. He ate breakfast, which is a good sign (raw bacon and wet food).

I leave for a couple of hours to have lunch with my parents and pick Sage up from school and take her to the orthodontist and then to the grocery store. When we return home at about 4:30, I know right away that something isn't right. It doesn't smell good. I don't see anything in the entryway or in the kitchen, as Maya and Topaz dance around us in greeting. As I step into the living room, I see Ody. He is crammed in the corner of the room, under the piano, squished between a piano leg and the side of the wooden frame of the dog bed. I know before I go any closer that he is sitting in his own feces. His hind legs are splayed out straight, and his front legs are sliding on the wood floor as he struggles to keep himself in sitting position. I have to crawl into the corner, stick my hand under his rear (in the feces) and push his butt up. Once standing, he staggers off to the back door, backside covered with greenish brown mash.

Sage runs in from outside and stops short. "Oh, ewww!!!" she says and runs back outside. She does her homework on the front porch.

I'm not sure where to start. Leaving groceries in the car, I change into old clothes, fill the bathtub and fetch Ody from outside. I wipe off as much poop as I can with paper towels and then lift him into the tub. My back is out of whack this week, so lifting and carrying all sixty-five pounds hurts. I scrub and rinse, scrub and rinse, until the water finally runs clear. I rub Ody with towels, and reassure him that he is a good dog.

I take a break to call the vet. Then I clean the wood and the carpet under the piano. The whole house smells. When Sage finally comes in she stands with crinkled nose and says, "It smells like a nursing home in here."

I ask the vet if (a)I should try giving Ody some antianxiety medication, so that he won't feel so distressed by not being able to stand or walk and (b)if I should try putting Ody in diapers. On the diapers, she says no. And on the antianxiety, she also says no. Her suggestion is that we try dosing Ody with steroid, even though there is a possibility that it will make him quite sick. At this point, we're down to very few options. We've avoided steroid because of his elevated liver enzymes, but given how severe the neurological impairment is, she thinks we should go ahead and give it a try. We'll know within a day or two if the steroid helps and, also, if it is going to make him sicker. I'm reluctant, so I tell her I'll think about it.

She also says to try acupuncture and canine physical therapy. These, she admits, are last ditch efforts. She also seems to want me to have Ody x-rayed, to look at his spine and see if there is an impingement. If there is, then back surgery would be in order but, she says, "I don't think I would recommend that, at his age." Why get an x-ray then, I wonder to myself. But I say to the vet, "I'll think about it."

Although I'm beginning to think more about the end of Ody's life, the idea of euthanizing him, at this point, still makes me shudder. I think the reason may be that he is clearly not actively dying, yet. He is still very much alive. He goes on walks, sniffs the air, marks his territory, and loves those hotdogs. I suspect things will change, in my mind, when Ody can no longer stand up, which may not be more than a few days away, at the rate things seem to be going.

NOVEMBER 8, 2010

Chris said he found Ody in the middle of the night, all tangled up in the elliptical trainer. His legs were caught between the steps. How he even found his way back into that far corner of the living room, I cannot imagine. The dogs never go back there, any of them. The only thing I can think is that Ody got confused about where he was. Chris heard the rattling sounds—it was enough to wake him—and came to investigate. Luckily! If Ody had fallen over while his legs were caught, he would have broken some bones.

Had another talk with the vet yesterday and told her I would try giving Ody steroids. The risks are starting to seem worth it. She said fine, and said she would have the prescription ready and that it would be waiting in the drop box outside her Berthoud office. I found myself too busy on Saturday to drive the twenty to thirty minutes to pick up the medicine. I didn't go today either, though I could have taken some time away from writing. And I find myself thinking about why. I am obviously resisting. I'm scared of the steroids—I don't want to hurt Ody.

I am feeling this same ambivalence about the canine rehabilitation place that the vet keeps telling me to go to. I made an appointment for next week, but I find myself dragging my feet. It will be very stressful for Ody to load up in the car and go to a strange place, and I find myself wondering if there really could be any benefit; it seems like tinkering with the sinking Titanic. Increased mobility would be a huge benefit for Ody, but given the state of his body and mind, I just don't see the therapy doing more good than harm. I know I shouldn't conclude this without at least giving it a try, so I will probably take Ody to the appointment.

Chris thinks it is a good idea to try the steroid, despite the risk.

NOVEMBER 10, 2010

I finally started Ody on the prednisolone last night. Miraculously, he is actually standing up straight today. His hind end looks like a normal dog's hind end. I worry about his liver, though . . . I keep watching for signs that something isn't right (not sure what signs, exactly).

NOVEMBER 12, 2010

Day 4 of Ody's prednisolone. He is better, yes. Definitely more mobile. And, he is driving me nuts. He has all this energy now and he's using it to follow me around and pant on me. All day he's been right there, right behind me, right next to me, panting. Chris said that steroids can make you feel ravenously hungry. Maybe all this following me around means that Ody is starving. Given how thin he is, I'm feeding him pretty much whatever he wants. Chris also says steroids can make you feel really "jacked up."

NOVEMBER 16, 2010

Ody is driving me crazy! All day long, as I work at the computer, practice the piano, cook, read—he follows me around and pants. He just looks at me and pants. Always he points his muzzle right at my face so I get the full effect of his bad breath. What does he want? Need? Is he in distress? Is this the steroid talking?

I've gradually lowered the dosage of the steroid over the past week, as instructed. The less Ody gets, the more he backslides into immobility. The lower dose still seems better than nothing, but the miracle has dissipated. The good news is that he shows no obvious signs of liver failure. But nothing I do seems to comfort his restlessness.

Ody's bladder is starting to leak a little. I've noticed wet spots on the couch, just in that area, when he's sleeping. Good thing for couch covers.

NOVEMBER 18, 2010

I think Ody is now quite blind. He is certainly almost completely deaf. Yesterday I called out from two feet behind him and he didn't turn. I raised my voice and he still didn't hear. Finally, I clapped my hands together and he turned.

His nose is permanently rough and cracked now.

Chris speaks of the prednisolone as hospice care. That's what he said when I told him I worried about the effects on Ody's liver. This is comfort care now. We'll do what makes him most enjoy life, even if it shortens his life a bit. We have stepped onto a down escalator from which there is no retreat.

I had the unfortunate experience of stepping in another fresh poop in the middle of the night. I was sleeping in the basement because Chris had a cold and was snoring. I heard Ody overhead, restless and wandering back and forth. I figured that at any moment he would begin barking to go outside, so I decided to just get it over with. I climbed out of my warm covers and stumbled upstairs. Ody wasn't right by the back door, as he usually is when ready to go outside. But I heard him in the living room, so I stepped into the kitchen and there it was: that warm, soft, oozing feeling . . . it hit my nose just milliseconds before my foot went down, but it was too late. There was one big (now squished) pile and a few small dots of poop. Funny thing was I actually felt relieved. This time Ody hadn't yet stepped in it and tracked it all over the entire house, so the cleanup was relatively easy. There is something about dog poop, though, that permeates your skin and even after several hand washings I couldn't quite rid myself of the smell, so as I crawled back into bed and tried to sleep I was surrounded by a faint unpleasant odor.

NOVEMBER 21, 2010

We came home from a soccer game in the evening, and that unmistakable smell filled the house. Ody had pooped on the throw rugs in the living room and had then stepped in it and tracked it all over the floor and all over his waterbed and his oatmeal dog-bone bed under the piano. I think we're going to need to confine him to a small space while we're gone— maybe in the kitchen? Or, just not leave for very long. We were gone about four hours. That must be too long.

Even though we clean everywhere we can see poop, and everywhere we smell it, I don't think we could have found every trace. I live in Poo-Ville.

We decided (Chris suggested and I agree) to take Ody off the predniso- lone. He seems miserable, and besides this, he is absolutely and completely driving us bananas. He is everywhere we are and always panting. If you are sitting on the couch eating a snack, he is right there, a foot from your face, panting. And he stands right in the middle of the kitchen. I feel bad shooing him out, but I would never let Maya or Topaz just stand there . . . why should I let Ody have special treatment? I guess I do give him special treatment, but I shouldn't assume that since he's old he no longer has to live by the same rules as other dogs in the house. When I try to scoot Ody out of the kitchen he tries to bite me. He is ornery these days. He won't move if I simply tell him "out" and point my finger (though all the dogs, including Ody, know this command); so I try to coax him along with my hands. He sort of jumps and turns—stiffly, with peg legs—shows me the whites of his eyes, and tries to get my arm in his jaws. He'll trot into the living room, and then as soon as I turn back to cooking, he'll be right in the middle of the kitchen, panting. I grumble under my breath, "Ody, stop your breathing."

I get different messages from the vet (who truthfully I'm not sure I com- pletely trust) and Chris, who is not a vet but whose training in human med- icine gives him good insight. The vet says to take Ody to the rehab place, and that she especially thinks that laser treatment will help. Chris says try one therapy at a time (steroid first; then rehab), so you know what, exactly, is helping, if anything. He also, when I mention laser treatment, assumes his skeptical expression. "It'll be a waste of money," he says.

NOVEMBER 24, 2010

Ody has been off the steroid for one full day and two nights. Already he seems much less restless, and he has stopped the incessant panting. He still pants, as usual, but not 24–7. Unfortunately, he also seems back to his old stiffness and sagging behind. Ody takes "sagging butt" to a whole new level.

THANKSGIVING, 2010

We took Ody for a hike today up at Left Hand Canyon. We had to carry him up the little hill at the start, but after that the trail is flat and easy. He seemed to enjoy himself.

NOVEMBER 28, 2010.

Sometime this afternoon, Ody began falling. It started with another episode of getting trapped behind the piano, his back legs out stiff, as if in a muscle cramp. Then he fell in the middle of the living room floor, then in the bathroom. He has fallen at least ten times in the last several hours and is very anxious, panting and sort of inching around after us. His gaze stays fixed on me, as if he's asking, "What's happening to me?"

Chris says that if Ody keeps falling like this, over and over, that by Wednesday we will need to make a decision to put him down. He is suffering, Chris says. I can't reply; I feel myself pushing the thought away. Just two days ago I gave Ody a score of 42 on the Pawspice scale—7 points above euthanasia. He couldn't decline that quickly. I think, instead, that I will take Ody to the canine rehab place and they will have a good solution to his problem. Maybe some magic bullet. I know, too, that I won't call the rehab clinic in the morning. It seems cruel to Ody to prolong this, to pretend that his problems are fixable. I follow him around all afternoon, picking him up, until finally he settles onto his waterbed and falls asleep.

At dinner, he gobbles up a bowl of fresh hamburger and rice. This makes me feel better. Food is a barometer of desire to live, isn't it? Tomorrow will be a better day.

6

Blue Needle

Animal hospice does a great service by encouraging us not to end life abruptly, if the dying process can be achieved without intense or protracted suffering. It also promotes a keen attention to pain and palliation. But when things become very grim for our animals, we also have the option of helping them die quickly and with relatively little pain—and this, it seems to me, is a very good thing. A good death can take many forms; we should embrace the possibilities.

Yet the terrain of animal euthanasia is fraught with traps and dangers. At its best, euthanasia can be an affirmation that we care deeply about and respect the life over which we've taken responsibility. Unfortunately, it can be many other things as well. It can be a tool for killing; it can be a way of shirking responsibility; it can be a weapon of cruelty and a reflection of how little we value the lives of animals.

THE VOCABULARY OF KILLING; OR, SAYING WHAT WE MEAN

Euthanasia means, literally, a good or easy death (from the Greek *eu* "good, well" + *thanatos* "death"). According to Webster's, to euthanize (or euthanatize) is to kill intentionally a hopelessly sick or injured individual, in as painless a way as possible, for the sake of mercy. Euthanasia is used in a variety of contexts, and within each the meaning shifts—sometimes imperceptibly, sometimes quite radically. Look closely, and you will notice that the definitions fall into two broad categories. In the context of humane and painless killing of animals, the word has generally positive connotations. In relation to people, the word euthanasia is most often used to create an aura of danger.

Definitional confusion is endemic to the human end-of-life conversation. For example, people struggle (unsuccessfully) to distinguish between active and passive euthanasia and among voluntary, nonvoluntary, and involuntary euthanasia. Sometimes withdrawing or withholding life support is mistakenly called euthanasia or, as I have seen it more than once, "limited euthanasia." Opponents of palliative sedation refer to it as "slow euthanasia." And many who oppose physician-assisted suicide will call it euthanasia, though technically speaking it isn't. Bioethicists have been hard at work for several decades trying to clean up the end-of-life vocabulary because being careful with our language forces us to clarify intentions and motivations and helps us see the moral ripple effect of particular policies and procedures.

This same kind of careful ethical work should be applied to animal death. Yet the work will be different. While human bioethics suffers from too much complex vocabulary, animal bioethics suffers from too little. We have too few words to describe how animals die at our hands. Indeed, "euthanasia" has become a catchall to describe a wide variety of killing practices; the term covers too much ground and obscures too many moral nuances.

We need to be able to distinguish, with language, between those contexts within which "euthanasia" says what we mean and those within which it simply serves as a euphemism for killing animals who aren't ready to die. Many varieties of euthanasia are requested by pet owners and carried out by vets. Some of these are ethically appropriate, some of them not, and many of them in some gray area in between. Is there any way to use more nuanced language so that we can be more transparent about our actions and motivations? Perhaps modify the term when appropriate—as in, for example, "convenience euthanasia" and "premature euthanasia"?

This is somewhat trickier, but I would argue that we should also avoid calling the killing of healthy animals in shelters "euthanasia," even if we judge that we are, in the broad scheme of things, doing these animals a favor. My reasoning is that killing healthy animals really serves human purposes and is not in the best interests of the animals themselves. Veterinarians and advocates for animals in shelters may disagree with me on this point. Several have told me that as long as the process of death is good—if we can achieve, for these animals, a death that is free of distress and pain—then it should be called "euthanasia," regardless of the reason behind the death. I'm not sure what the right answer is, but a careful discussion about language is certainly worth having.

SOME MORAL INGREDIENTS

When we take dogs and other companion animals into our lives and homes, we accept responsibility to care for them throughout their lifetime. We control many aspects of their lives—when and what they eat, when and where they eliminate, where they live, when they will be inside and when outside, whether they are allowed to engage in sex and with whom, whether and with whom they may produce offspring, and often when and how they die. But what exactly does taking responsibility for the end mean? Are there ever compelling reasons to actively end the life of an animal? Could it ever be a good decision to end Ody's life? It should be clear by now that I think euthanasia can be the right thing to do. But I also think it is performed far too often, too early, and for the wrong reasons. It is worth thinking carefully about the conditions under which the act of euthanasia is morally appropriate.

WHAT

What we are doing, of course, is trying to identify some of the considerations that might go into the decision to euthanize an animal.

WHY

Euthanasia is an appropriate choice if an animal is clearly suffering and if there is no feasible way to ease this suffering. Keeping an animal comfortable is the ultimate goal, and when we run out of ways to relieve discomfort, euthanasia may be our best option. Jerrold Tannenbaum, a professor of veterinary ethics at the University of California Davis, suggests the following as a case of clearly justified euthanasia: (1) the patient is suffering from a condition for which veterinary medicine cannot offer a cure or solution at any cost; (2) the condition is already causing the animal severe pain for which palliative measures are not available; and (3) the human client is psychologically able to make a voluntary and rational decision. This can serve as a starting point for reasoning about more difficult cases.

In every instance of euthanasia, we would do well to begin with the question, "What is in this particular animal's best interests?" And we must work very hard to make sure that we've done the best job of interpretation we can. Much of what we talked about in the last chapter, on QOL, is called into play here. You weigh good and bad; you do the pawspice scale or some version of McMillan's QOL calculation. Is there pain? How severe? Does my animal still experience moments of joy? Do bad days outnumber good? What do others say—the vet? My friends? Other family members?

WHEN

It has always struck me as odd to choose the moment of your animal's death, to make an appointment to have them killed. Yet this is exactly what we often do. For Ody, the appointment was at 6:30 on November 29, 2010.

The issue of timing is one of the most excruciating for pet owners. You often hear "They will tell you when its time" or "You will know when it is time." But, really? I don't buy this. You don't know and they can't tell you—and no matter what, you will agonize over whether it is too soon or too late. You will never know, not before and not after the fact. You will be left wondering whether you were wrong to set a time for death, and whether this awesome responsibility shouldn't rest always in the hands of some greater power, someone who doesn't suffer from the same limitations and blind spots.

I think we need to set aside the notion that there is a Right Time—some moral target that we need to hit precisely. Our goal is not to pinpoint but to find a golden mean between too soon and too late, between premature and overdue. Working with a veterinarian, we can seek to understand our animal's illness or injury, to know what kind of deterioration or changes they might experience. Particularly if we shift into a hospice mindset, we can define treatment goals (using a kind of advanced directive: What does your pet value? What do you value?); we can decide what the options are (different kinds of treatment, palliative care, benefits/burdens of each option, as much as these can be discerned); we can weigh quality of time left against quantity of time left.

Perhaps there is a watershed event that shifts the balance, when an animal crosses some invisible boundary into suffering and into a realm of "anytime now would be good." For Ody, this was the day that he could no longer stand up, the day he fell over and over and over and had to simply wait, wherever he was, for someone to pick him up. As I look back on my journal entries, I see a long windup; he had been on his trek into the realm of suffering for some time. Still, am I certain about my timing? Not at all.

A common refrain in the pet end-of-life literature is this: too soon is far better than too late. And as a popular saying in veterinary medicine goes, "I'd rather help my friend a month too soon than an hour too late." This is because "too late" can be really awful for the animal.

Why do I balk at the assurance that "They will let you know"? Because this places responsibility on the animal and removes responsibility from us. Our animal may indeed give us signs that they are suffering (refusing to eat, withdrawing into themselves), but it is we who must read the signs— and as should be abundantly clear from all that has come before here, the signs can be obscure. It is not easy to interpret pain in animals; we don't know their behavioral language, unless we undertake the sustained work necessary to understand it. And our interpretation of their "signs" is, as often as not, clouded by our own interests, presuppositions, and ignorance. And, yes, by our love for them.

I asked one of our local vets whether he thinks people wait too long. "Yes, very much so," he answered. "Sometimes you just have to say to an owner that they are being selfish and that they need to let the animal go. Most people, once you point out how much pain an animal is in, will comply." He also told me that the number of requests for euthanasia spikes just prior to the holidays and at the first of the year when, he speculated, people might say to themselves, "We've been holding on too long; time to start the new year fresh."

WHERE

When I started working on this book, I assumed that if you chose euthanasia for your pet, you would, by necessity, go to a veterinarian's office. I worried about this, long before Ody was seriously ill. Ody has always hated the vet's office and would go into paroxysms of anxiety during a wellness visit or even if he so much as walked within a hundred yards of a veterinary office. I hated the idea of his final moments being filled with the anxiety of being in a place that smelled so bad. But I knew of no alternatives. Then I read somewhere that vets will often come out to the parking lot and euthanize an animal in the back of your car. This seemed much better—Ody loves the car. But still, a parking lot is not my image of an aesthetically pleasing and peaceful place for death. I think of our vet's office, sitting at the intersection of two busy roads, across the street from Curt's Cleaners and next to the shabby Quality Liquor store. Not nice.

Then someone mentioned to a vet who would come to your home and do the euthanasia there. After some searching in the local directory I found her: Dr. Kathleen Cooney, of Home to Heaven In-home Euthanasia Services. I know it sounds weird to be excited about euthanasia, but I can't tell you how relieved I was to know that my options—Ody's options—were expanding. Home seemed like just the place.

Home-to-Heaven offers round-the-clock and emergency euthanasia services, 365 days a year. Dr. Cooney is a full-time euthanasia vet. And she and her team of veterinarians are very busy. Home-to-Heaven responds to about thirty requests per week (for an area in northern Colorado and southeastern Wyoming, maybe a hundred or hundred and fifty square miles). Her practice has been steadily growing since she set up shop a few years ago, as more people become aware of the option of home euthanasia. Cooney openly embraces death as an outcome, and doesn't see her work as, in any way, dirty; she is very passionate about euthanasia as a gift of release that we offer our animals. She was my go-to person for questions about euthanasia, and I see in her work a vision of euthanasia's future possibilities.

Most of Dr. Cooney's euthanasias take place in the home, which she says is the very best location. And almost all are done same day—what she would consider emergency euthanasia. She does a lot of middle-of-the-night calls, which she actually likes because it is quiet and because people really appreciate your presence. She charges between $160 and $250, depending on how far she has to travel. (This does not include the cost of disposing of the body.) This is no more expensive than a veterinary office visit, and much cheaper than an emergency hospital, which can be two or three times as much. Incidentally, some pets might feel more defensive and territorial at home and could be calmer at the vet's office. This is very personal and will be unique to each animal. Dr. Cooney's basic guideline is this: choose the place where your animal will be the most at peace and least stressed.

Dr. Cooney is also embarking upon an as-yet untried adventure: She has just opened The Euthanasia Center, the first of its kind for animals. It is an alternative to in-home euthanasia, which some people don't like or cannot afford, and an alternative to the veterinary clinic. When you perform euthanasia, she says, you need everything to slow down and be quiet—a state hard to achieve in a traditional full-service vet clinic. A scheduled euthanasia can be really disruptive to the flow of the day, and it can be near impossible to create a quiet and peaceful space for the animal and its human family. The Euthanasia Center will be even less expensive than an in-home euthanasia and will make the service financially accessible to a greater range of clients.

I visited the Euthanasia Center on the outskirts of Loveland, Colorado, just before its official open house. It sits on an expanse of flat farmland, with a view of the Rocky Mountains to the northwest. The center is located in what used to be a detached garage next to Dr. Cooney's house and is purposefully designed to look and feel like a home. The euthanasia room is like a small, neat parlor, with lots of natural light. The color scheme is blue and tan, with a sofa and loveseat and a small bookshelf. Notably absent are a metal table and steel cabinets filled with medical supplies. And it smells normal—like a house, not like antiseptic cleaner. There is a pretty, patterned dog bed in the center of the carpet, for large dogs. Smaller animals, if the family and animal prefer, can be euthanized on the couch. There is a fenced yard out back, so that other family pets can be brought along.

WHO

Who should perform the euthanasia? That's an easy one: a well-trained and compassionate veterinarian.

Who should be present at euthanasia procedure? I've talked to many people who say that they dropped their animal off at the vet or shelter and then left. "I just couldn't deal with it," they say. Or, "It was too sad." This seems really unfair. We shouldn't abandon our animals at the end, simply because it is hard for us. Then again, it may sometimes make sense for a person not to witness the actual euthanasia procedure. Some people know that they will become overwrought, even hysterical, watching their animal die. Our animals have a keen sense of empathy and will be attuned to these emotions and may become concerned and anxious. We don't want our animal to die in a state of distress; we want it to be peaceful. So having other loving and caring but less emotionally invested individuals present might make sense in some situations.

Should we bring friends and family? There is no reason not to, although it is important that the actual procedure be a calm and quiet time for the animal. Experts on bereavement disagree about whether children should watch a pet being euthanized, but the weight of opinion weighs more heavily toward the positive value of having children participate in the final passing of a beloved pet. This is obviously quite personal and depends a great deal on particular circumstances and on the maturity of the child.

Should other animals be present for the euthanasia? I've come across different answers to this question from veterinarians and behaviorists. Some say no: other animals should not witness the death because it will be traumatic for them. But they should definitely be present for the wake: they need to see and smell the body.

Others say yes, other animals should absolutely be present for the entire event. It is beneficial for the animal being euthanized to have their companions present, and it is of value to the remaining animals because they may well understand what is happening. I asked Dr. Cooney if other animals in the house are usually present during the euthanasias she performs. Often, she said, they are invited to be there, unless they are really disruptive. I also asked if she thinks the other animals are aware of what's happening, to which she said, "Sometimes yes, sometimes no." She doesn't think that animals understand death, per se, but they may be aware of some change. She told two stories. In one house, she put down a yellow lab. Her two lab companions were in the other room when the animal died, and just at the moment of her death, they both began to howl. Labs, apparently, don't usually howl, and these two never had. When they were let into the room, they ran to the body and stood over it. In another house, she put a cat to sleep on one of the beds. There were three other animals in the house— two cats and a dog. They were not present for the euthanasia, but later, after the body had been removed, all three went into the bedroom, got on the bed, and curled up in a circle, with the dead cat's spot in the middle.

HOW

Skillful technique and appropriate choice of euthanizing agent can make all the difference between a good death and a death that is unnecessarily prolonged and stressful. I read through Dr. Cooney's forthcoming book, *In-Home Pet Euthanasia Techniques*, which she wrote to help veterinarians improve their euthanasia skills, and I was amazed at the complexity of the procedure. It is far from a simple injection, but requires a detailed understanding of drug dosages, the pros and cons of various types of injections, how certain health conditions can alter what you need to do, and how all these complex veterinary decisions must take place while at the same time being aware of how your work will be viewed by the pet's highly aroused and emotionally distraught human family. As Dr. Cooney notes throughout her book, there are various ways in which the procedure can go wrong—from running out of euthanasia solution, to missing the heart during an intracardiac injection, to puncturing through a vein—and one needs to know how to avoid these pitfalls and be prepared for the unexpected.

Euthanizing agents cause death by three basic mechanisms: (1) direct or indirect hypoxia (the body, or some part of the body like the brain, is deprived of oxygen); (2) direct depression of neurons necessary for life function; and (3) physical disruption of brain activity and destruction of neurons necessary for life. The American Veterinary Medical Association offers a detailed examination of twenty-four acceptable methods of killing and seventeen unacceptable methods. Among acceptable methods, we have the inhalant agents (inhalant anesthetics, carbon dioxide, nitrogen, argon, carbon monoxide), noninhalant pharmaceutical agents (barbituric acid derivatives, pentobarbital combinations, chloral hydrate, T-61, tricaine methane sulfonate, potassium chloride in conjunction with prior general anesthesia), and physical methods (penetrating captive bolt, blow to the head, gunshot, cervical dislocation, decapitation, electrocution, microwave irradiation, thoracic compression, kill traps, maceration, exsanguination, stunning, and pithing).

The use of injectable euthanasia agents (technically speaking the noninhalant pharmaceutical agents) is considered by the AVMA to be the most reliable, rapid, and humane method, "when it can be performed without causing fear or distress in the animal." If you go to a vet to euthanize your companion animal, or have a vet come to you, this is the technique they will use. This technique is usually referred to as euthanasia by injection (EBI) rather than lethal injection. Most vets use barbituric acid derivatives such as sodium pentobarbital or pentobarbital combinations. Barbiturates are the class of drugs used to induce anesthesia in humans for surgery. They cause loss of consciousness and loss of pain sensation and, in higher doses, will suppress the cardiovascular and respiratory systems. So, the animal loses consciousness and a few moments later the heart and lungs cease functioning. Dr. Cooney always gives an animal a pre-euthanasia sedative such as xylazine, unless there is good medical reason not to. Although the use of pre-euthanasia sedatives is not universal among vets, Dr. Cooney believes it should become an integral part of the euthanasia procedure because it allows for a more peaceful passing for the animal and an easier time for the family.

In addition to being the name of an obscure rock band, Fatal-Plus is the name of a commonly used euthanasia solution. Other marketed solutions include Euthasol, Sleepaway, Beuthanasia-D, Socumb-6, Repose, and Somlethal. I find these names fascinating, in a creepy kind of way.

Fatal-Plus was the solution used in Ody's needle. Because it is pure sodium pentobarbital, Fatal-Plus is a Schedule II Controlled Substance and is regulated by the Drug Enforcement Agency. Some combination formulations, such as Beuthansia-D (pentobarbital sodium combined with phenytoin sodium), are Schedule III, and slightly less regulated than Schedule II drugs because they are less easy to abuse. A veterinary license is generally required to buy these solutions. Shelters buy large quantities of euthanasia drugs, and many states recognize that euthanasia is an essential part of running a shelter and allow for purchase and use without a vet. Fatal-Plus is, as one vet told me, "dirt cheap"—thirty-three cents a milliliter. (For Ody's injection, the cost would be about two dollars.) Euthanasia formulations generally contain a pink or blue dye, to distinguish them from drugs intended for therapeutic use. After all those years of threatening Ody with the pink needle if he ate one more couch or stole one more plate of cookies off the counter, the fluid in his syringe turned out to be blue.

When you buy a prescription drug, it comes with a detailed factsheet about active and inactive ingredients, drug actions, side effects, and contraindications. Here's the fact sheet for Fatal-Plus:

Fatal-Plus®
Indications: For fast and humane euthanasia of all animals, regardless of species.
Administer 1 ml of solution per 10 pounds of body weight.
Active ingredient: Pentobarbital sodium.
Actions: Produces classic euthanasia by sequentially depressing the cerebral cortex, the lungs and the heart. Action on target organs gives

humane euthanasia of unparalleled speed, effectiveness and specificity. Instant unconsciousness is induced with simultaneous collapse of the animal. Deep pentobarbital anesthesia ensues with blood pressure fall, stoppage of breathing and cerebral death. Cardiac function stops, quickly and irreversibly.

A source told me about a new drug under development that might, were it approved, be better than sodium pentobarbital formulations. It is a propofol combination drug. (Propofol may sound familiar: it is the drug administered by Dr. Conrad Murray to Michael Jackson that caused the pop star's death.) It would cost about fifteen cents more per milliliter than Fatal-Plus, but the higher cost would be well worth it (said my secret source). "Every vet in the country would use it." Propofol is a hypnotic agent and has some recreational appeal, but because the propofol is mixed with other drugs and cannot be isolated, its only conceivable use would be for euthanasia. This would remove the potential for abuse, thus allowing the solution to be less heavily regulated than Fatal-Plus and other barbiturate drugs. The absence of barbiturates also means that the drug would potentially be safe for use in food animals, though careful studies on drug residues in carcasses have not been conducted. (Wildlife or other animals who feed on carcasses containing barbiturates can die from them—a serious drawback to sodium pentobarbital.)

My source was most excited about the fact that the propofol combination drug would completely remove the "side effects" of euthanasia. My mind raced around trying to figure out what possible side effects euthanasia might have, but I finally just had to ask. Side effects are what vets call the twitchings, agonal gasps, and other signs of death that can upset a family watching their animal being euthanized. The ideal euthanasia solution would make death mimic sleep.

WHAT EXACTLY IS "DEAD"?

Which brings us to some uncomfortable questions. What exactly does death look like? Does it look like going to sleep? And is it always clear when something or somebody is dead?

Death isn't a straightforward proposition. Technically speaking, death is the termination of the various biological functions that sustain life.

Yet biological functions shut down gradually. Death is a physiological process—sometimes drawn out over an extended period.

Because death is a physiological process, there are various points during this process at which we might decide that death has, officially, occurred. For example, does death occur when respiration ceases, or when the brain shuts down? With people, at least, when exactly a person becomes dead is highly contentious. The lack of consensus stems partly from the fact that medical advances keep changing our ideas about what it means to be dead. The traditional definition of death as cessation of respiration began shifting during the 1960s and 1970s toward a definition that centers on brain function. A person's body can be performing basic functions, like breathing, with the help of machines, but if there is no brain activity the person can be considered, and declared, dead. The impetus for this change in legal definition of death was partly philosophical—a growing consensus that consciousness, not a corporeal body, was the essential ingredient of "personhood"—and partly practical—the need for transplantable organs. Yet even brain death is not so obvious. Scholars still argue over whether death should be defined the irreversible cessation of electrical activity in the brain ("brain death")—the current legal definition in most states—or the cessation of functioning in neocortex ("higher brain death"), since the neocortex is believed to be the seat of personality and thought.

Has there been any comparable controversy over the definition or conceptualization of animal death? As far as I know, there has not. People just don't fret about animal death as they do human death, partly because animals don't matter that much and partly because animals have generally been denied such philosophically challenging capacities as "personhood." Nor, until very recently, were animals subjected to advanced life-prolonging medical treatments such as artificial respirators and percutaneous endoscopic gastrostomy tubes that can make death ambiguous (except, of course, as the experimental bedrock on which such technologies were developed for people).

Putting aside conceptual ambiguities, isn't it obvious when a person or animal is really, truly dead? Actually, no. We can poke at a body and watch for movement, but we have to wait for rigor mortis or even the odor of decay in order to be certain. Not just anybody can pronounce a human being dead. In fact, death requires medical diagnosis by an MD or, depending on the jurisdiction and the details of the death, a paramedic or a registered nurse. Even so, medical professionals occasionally make mistakes, declaring people dead who really aren't. If a person is hypothermic or has taken barbiturates, they may appear to be dead on medical examination. Cases have been reported of people waking up in the morgue or on the embalmer's table, and it is a deep fear of being mistaken for dead that drives the invention and sale of caskets with built-in alarm systems.

What about animals? Is death obvious? Apparently not, and as a case in point consider the story of Mia, a ten-year-old Rottweiler. Mia's family, after watching her suffer from a crippling arthritis, decided to have her euthanized. The vet administered the standard euthanasia protocol, and Mia's mourning owner took her dead body home and placed it in the garage, to be buried the next day. Imagine the owner's shock when Mia greeted him at the garage door in the morning.

A great deal of hand-wringing took place over this case within Internet pet discussion groups, and there followed many reports of similar, not-so-effectively euthanized dogs. How do we know this doesn't happen all the time? We don't. Animals don't receive death certificates. An animal can be declared dead by anyone, no matter how ignorant of biology, and with or without adequate evidence. If animals that are euthanized are then placed in a freezer until the crematory or rendering truck comes to pick up their bodies, how would we ever know for sure that the euthanasia had been complete? Veterinarian and journalist Patty Kuhly recommends the following checklist to make sure an animal is really, truly dead:

1. Absence of a pulse (by manual palpation).
2. Absence of a heartbeat (via stethoscope).
3. The absence of respiratory movements.
4. A change in coloration of the gums from pink to grayish (pallor mortis)
5. The onset of rigor mortis, in which the limbs become stiff (which can take between ten minutes and several hours).

We could add several more fail-safe indicators: algor mortis, the gradual cooling of the body following death; livor mortis, a settling of blood in the lower part of the body; and, of course, decomposition and the accompanying smell of decay. (I mean this quite seriously: it is part of our responsibility to ensure that death has actually occurred.)

One of the jobs of a euthanasia vet seeking to create a good death both for the animal and for the animal's family is to be aware of the natural stages of death following lethal injection and prepare the family so that they understand what is happening to their animal. Kathy Cooney describes the death process in her guide to euthanasia technique, alerting practitioners as to how death will likely occur after the administration of a barbiturate. The physical signs of death, although universal to all animals (including humans), can manifest differently from one animal to the next, depending on the animal's underlying health, whether a sedative is given, and according to method of euthanasia.

Cerebral death occurs almost immediately after the lethal injection is given. At this stage, an animal's body may show "lifelike" physical movements such as muscle twitching and stretching of the legs. The second stage of death is respiratory arrest, when the animal stops breathing. During this phase, an animal may shift from slow, rhythmic breathing to a quick volley of rapid inhalations, followed by a complete cessation of breathing. Even though the breathing centers within the brain have ceased functioning, the animal may still take some agonal breaths. These look like sharp irregular inhalations punctuated by long pauses. Cooney warns practitioners that the pet may also open its mouth wide, "make a 'gasping' sound, and curl its body with each breath." These agonal breaths can make it look like the animal is suffocating, so the family needs to be reassured that these are purely reflexive and that the animal is completely unaware of what is happening.

The final stage of death is cardiac arrest. Usually, Cooney explains, the heart will stop beating very quickly—within about thirty seconds of administration. Cardiac electrical activity, she continues, "can occur for up to 28 minutes after euthanasia, even though no physical signs of life remain." When you take out your stethoscope and listen to the heart, "you are safe in saying the pet has died even if you hear faint fluttering within the chest." This is the sound of the final electrical impulses. However, she cautions, "if you still hear regular heartbeats 90 seconds after an intravenous or intracardiac injection, something is not right." The animal is not dead. You need, she says, to check the injection site to make sure the entire dose went into the vein, and if it didn't, give another full dose.

We come, finally, to something called postmortem side effects. Cooney outlines some of the most common side effects of death: curling of the tail, defecation or urination, opening of the eyelids, twitches of the whiskers or feet, tail hair fluffing, stretching of the legs, back, and neck, and muscle fasciculation (small, involuntary muscle contractions and relaxations). These changes occur in the several minutes after death and can make us worry—if we are not prepared—that our animal is still alive.

CONVENIENCE EUTHANASIA

Not many years ago, vets would routinely and without question put down healthy animals at a human client's request. No one differentiated between motivations for euthanasia—it was all one, and not taken all that seriously, from a moral point of view. Now, many vets balk at euthanizing healthy animals. There is even a term that designates these morally suspect requests.

"Convenience euthanasia" describes the killing of a healthy pet, at the owner's request, for the sake of the owner's convenience. You typically hear this as a phrase of disapprobation. Bernie Rollin offers a few examples in his *Introduction to Veterinary Medical Ethics* of what would clearly be called convenience euthanasia. A woman with healthy five-year-old cocker spaniel is moving and can't take the dog to her new apartment and her boyfriend doesn't like the dog; she asks a vet to euthanize. A woman brings in a five-year-old male cat who started spraying after the birth of the woman's baby. She asks vet to destroy the cat. A breeder brings in a healthy six-week-old puppy with a moderate overbite. Because the dog is not show quality, the breeder asks the vet to euthanize.

How often does convenience euthanasia occur? I couldn't put my finger on any statistics, so all I can say is that every vet I talked to says that it happens and could easily call to mind a good many examples of euthanasia requests that had made them uncomfortable. We could certainly find veterinarians around the country for whom these requests present no particular problem. Some vets work under the assumption that whatever a client (human) asks them to do, they do, as long as it's legal. So, tail docking, declawing, ear cropping, and convenience euthanasia—it is all part of a day's work. A few vets, at the other extreme, simply refuse outright to perform euthanasia in situations such as these, no matter the business consequences.

But most that I've talked to find the moral landscape extremely murky. They are loath to perform convenience euthanasia but often do so anyway because they believe it is in the best interests of the animal. The first response is often to try to talk the client out of killing their animal. But when this fails, the vet will often comply. The reasoning goes like this: either the owner will try to do it themselves (not a good option), will relinquish the animal to a shelter (where it may, after several stressful and awful days, be killed—well or poorly, also not a good option), or will simply move on down the road until they find a willing vet (thereby simply prolonging the process). By agreeing, at least the vet can assure that the process is peaceful and quick for the animal.

BORDERLINE CASES

Convenience euthanasias are usually pretty clearly not in the best interests of the animal. But consider the more troublesome cases, where we cannot say for certain that an animal is in great suffering and that death is just around the corner but where the owner is nevertheless requesting euthanasia for the sake of the animal. Every vet I talked to spoke with emotion about these borderline cases, and each of them had a number of such experiences to relate. Kathy Cooney writes, "If you provide this service long enough, you will face a family's request to euthanize a pet that you feel still has a good quality of life." What I heard over and over is that veterinarians almost always defer to the judgment of their clients and, as Dr. Cooney says, "try not to judge" (or not too much, anyway).

Dr. Cooney told me that she is rarely uncomfortable with people's requests, and she offered a few examples of requests that she ultimately felt were reasonable. Each of these gives me food for thought; they don't seem obviously right but not obviously wrong either, and I think, finally, that Dr. Cooney is probably correct: the animal's human is in the best position to make these decisions.

Sometimes an animal has a terminal diagnosis, and the family will ask to have him or her put down before they go away on vacation. To Dr. Cooney, this request makes sense: the animal will be stressed out when the family leaves and may die while they are gone or, at least, have an emergency come up. She also gave the example of a dog diagnosed with bone cancer. The family called the same day they were given the diagnosis and asked to have their dog put down. This request made sense to her. With bone cancer, the bones become very brittle. You might take the dog to the park or on a hike and his leg will just break. Now you've got an emergency; the dog is in pain, is frightened; and the euthanasia is performed under conditions of stress.

She once euthanized a cat with inappropriate urination. It was a healthy cat but simply would not stop peeing in the house, on the rug, on the furniture. The family tried everything they could think of: medical exams, pheromone spray, even moving to a new apartment. She euthanized this cat, reasoning that otherwise it would have gone to the shelter, which is already full, and would likely not be adopted because of its age and its behavioral issues. After months of stress in the shelter, it would be put down. She also euthanized a two-year-old boxer who suffered from profound anxiety. The dog's owner was herself in hospice and had about ten days to live. The dog was a wreck; all attempts to find a home had failed.

She also told me that she frequently euthanizes two animals together. For example, two dogs may be littermates who have been together their whole lives. Both are elderly, but maybe one is in worse shape than the other. Or, sometimes she'll even have a cat and dog who have been housemates for a long time and who are strongly bonded. A family may be concerned that the remaining pet will be too distressed at the loss of their companion and that it would be better to euthanize them at the same time.

What about this surprisingly common scenario: when a person dies, they specify in their will that they want their animal to be euthanized? They may feel that no one could care for their dear friend adequately or that their companion will be so bereft that death would be preferable to continued life. Or—and this is not unusual—a person simply wants their beloved pet to be buried in the same casket, so immediate death is required?

For all those who despair over the fate of animals in our world, we should take heart at the fact that there is now so much nuance within the world of animal euthanasia—that people distinguish, ethically, between good and bad motivations and agonize over borderline cases and that so much attention is given to perfecting humane methods. All this is very, very good.

BAD DOG!

Behavioral issues are thought to be the leading cause of dogs being relinquished to animal shelters and the reason that between 25 and 70 percent of dogs in shelters every year wind up under the needle. Among the euthanasias performed by veterinarians on companion animals, a large number are requested because of unresolved behavioral problems such as aggression or house soiling.

It is an odd habit of humans to deny generally that animals have intentions, yet when it comes to behavior that we find annoying or unacceptable, we anthropomorphize like crazy, ascribing to our dog all kinds of malicious or naughty thoughts. What better proof of this than the widespread reaction to chewed shoes or stolen-off-the-counter food: "Look at that dog! He knows he's guilty!" Dogs do not, according to the most recent research, have a sense of guilt, nor do they feel remorse. But they do have a keen sense of empathy. They cringe not because of regret over having eaten the steak off the counter but because we are angry. Ironically, behavioral issues are often the fault of the owner, who hasn't spent time and energy on training, who doesn't understand canine behavior, or who has unknowingly set his dog up for failure.

Fortunately, "bad dog" euthanasias are becoming less and less necessary as people better understand the causes of and solutions to behavioral problems. More people realize that dogs with behavioral issues are not bad to the bone but simply don't understand what we expect of them, usually because we haven't explained it in terms they can comprehend. New medications are available to treat psychological disorders such as separation anxiety, and new research is helping us understand the emotional and psychological nuances of behavior. A growing number of animal behaviorists are available to help us troubleshoot. Remember: behavioral problems are often a result of poor socializing and training, and pet owners need as much, if not more, socializing and training than do their dogs. Some behavioral problems—notably house soiling—can be traced to medical problems. And some behavioral problems stem from depression, boredom, anxiety—all of which may be treatable, through better care, more exercise and stimulation, and, when these fail, through medication.

DO-IT-YOURSELF EUTHANASIA

The euthanasia procedure sounds easy: a quick injection of a single drug. So why not just do it yourself? It seems appealing: you choose the time and place, and it's very inexpensive. I cannot verify how often people attempt euthanasia at home, but *USA Today* reported that the topic is being widely discussed on the Internet, and one can peruse various anecdotal accounts of the do-it-yourself variety.

The very first problem, of course, is that you can't use sodium phenobarbital at home, since is a controlled substance and can only be purchased by licensed veterinarians or animal shelters. I suppose it could be obtained illegally—most drugs can. But it seems likely that people would use other, more readily available, options. Several online discussions describe the use of "a drug that ends up stopping their heart." Not sure what this might be. Other possibilities come to mind as well (how many poisons do you have in your home at this moment?).

I can think of several reasons why someone might be driven to try do-it-yourself euthanasia—and most of these seem pretty flimsy, when weighed against the challenge of killing an animal well. A big one, of course, is money: it can cost a couple hundred dollars to take your dog or cat to the vet for euthanasia. Being a cheapskate is no excuse; but real financial constraints are a legitimate concern. Still, shelters usually offer euthanasia services at a reduced price, and sometimes even for free. My local shelter charges $40, and no appointment is necessary.

Others feel that ending their animal's life at home is more comfortable than going to a vet. They may believe that the trip to the vet would be painful or disturbing, or that dying in the vet's office would not be peaceful for the animals. This is a legitimate concern, but there are plenty of veterinarians who will come to the home to euthanize an animal. I suppose in-home services may never reach some rural corners of the country, but weigh the discomfort of euthanasia in a vet's office against the ugly possibilities of a botched job at home.

Yet another problem that arises for people is timing—their animal needs to be euthanized over the weekend or over a holiday, when veterinary offices and shelters are closed. One Internet story recounted the trials of a woman whose fifteen-year-old shih tzu deteriorated very rapidly over the period of a couple of days. The animal appeared to be in great pain, and the woman was anxious to have him euthanized immediately. She could find no vet within her area who would schedule the euthanasia that week, and she didn't want to go to an emergency clinic because of the high cost. She called her vet again and asked for pain medication, but the vet refused, saying that, given the dog's condition, pain medication could kill him (!). Feeling backed into a corner, she ended up doing it herself (which, incidentally, she found profoundly distressing).

I suppose there may be people who are simply independent and enterprising and fancy themselves knowledgeable in veterinary matters. I found detailed instructions on the Internet for euthanizing a small mammal using carbon dioxide. I imagine being a mad scientist: sticking Sage's rat Ninja in a big plastic jar or Ziploc bag, combining the baking soda and vinegar in a separate Tupperware, connecting the two containers with a piece of hosing, watching the chemical reaction while rubbing my hands together and cackling softly to myself. The website tells us that the AVMA approves the use of carbon dioxide for small animals weighing less than two pounds. Somehow I don't find this very comforting.

We should distinguish at-home euthanasia—where a person is trying to end the life of an ill, suffering animal for the sake of mercy—from the more common practice of at-home disposal of unwanted animals. Here the motivation is purely selfish: getting rid of something you don't want, with the least amount of trouble. For example, a friend of my father's calmly described how, after deciding to move into an assisted living facility, he took his two hound dogs out into the field on his farm and shot them with a rifle. Our local newspaper reported on a horse owner who apparently decided to dispose of his horse by a gunshot to the head. As we've seen, skill is required, and in this case, the bullet crushed the horse's face, but missed its brain. The person obviously didn't stay around to see whether the procedure had been effective; the horse was found in a ditch on the side of the road by passersby, staggering and bloody.

KILLING IN SHELTERS

Many people—animal lovers and activists among them—believe that killing animals in shelters is an act of mercy that precludes further suffering by the animal. Whether merciful or not, we should be cautious about using "euthanasia" as a blanket term to cover all the killing because this language obscures the reality of what's really happening. There are many varieties of death occurring in our nation's shelters and pounds, and many of these deaths are pretty awful. It would require vigorous stretching of the imagination to describe as a "good death" the experience of a frightened dog being snagged with a catch pole and shoved into a concrete oven with ten other terrified dogs and then gassed with carbon monoxide. Even if the majority of shelter killings were to be accomplished through injection, which can (under the right circumstances) be relatively stress free and painless, we should still think carefully about whether these ought to be labeled euthanasias. "Mercy killings" is perhaps more accurate.

We are talking here about a large number of animal deaths. According to the most recent data, about 1.5 million dogs and 1.8 million cats are euthanized each year in our nation's shelters, with the highest rates found in the Midwest and the south Atlantic. The International Institute for Animal Law reports wide disparity among shelters in their methods and application of euthanasia. Illinois, Michigan, North Carolina, and Texas still permit the use of a gas chamber, despite ample evidence that these deaths can be terrifying, painful, and protracted—sometimes it takes as long as thirty minutes for animals in a gas chamber to die. The institute notes that "problems stemming from inadequate training, insufficient funding, indifference to animal suffering, and failure to recognize the need to change and update procedures, are found everywhere, from small rural shelters to large city facilities." There is an urgent need for a consensus on humane methods of euthanasia and implementation of humane techniques.

Doug Fakkema, one of the nation's experts on shelter euthanasia, is a strong advocate of euthanasia by injection (EBI). Ideally, he says, all killings in shelters should be done through injection of a sodium pentobarbital solution. One of the reasons the use of a single drug method appeals to Fakkema, and to humane organizations and veterinarians, is that injection of sodium pentobarbital offers a procedure with minimal risk of pain, with a wide margin for error, and where the costs (to the animals) of mistakes can be minimized. Within the nation's shelters and pounds, euthanasia is rarely performed by a vet. Instead, the work of killing animals falls to low-paid shelter workers or animal control officers, who usually have no formal veterinary training. In some states, shelter workers who perform euthanasia are required to attend several hours of training to become a Certified Euthanasia Technician. In other states, no formal training is necessary. Fakkema would like to see a sixteen-hour certification course required by all states. Still, good technique isn't enough. The humaneness of shelter euthanasia depends a great deal on who performs the work. Even with EBI, says Fakkema, "the person doing it must be humane; the animal must be loved, petted, touched." He then adds, with sad irony, "Anybody who wants to do this work shouldn't do it."

Given the massive numbers of animals that must be dispatched by shelters (which are often underfunded and operating within severe budget constraints), cost is an important consideration. Gas chambers have long been thought to be cheaper and more efficient than EBI because you can kill a number of animals at once. And the gas itself is cheaper than sodium pentobarbital. Among the shelters that remain gas-chamber holdouts, tight budget is the typical excuse. To challenge this, Fakkema compiled a detailed cost analysis matrix for a municipal animal shelter in North Carolina. About fifteen animals are euthanized, on average, each day. Considering all potential costs associated with the procedure—equipment, personal, material, labor—the average cost per animal of euthanasia by carbon monoxide chamber is $2.77. This is without the use of a tranquilizer and with only one operator, both of which increase potential stress on the animals. Best case scenario, with tranquilizers and two operators, the cost rises to $4.98 per animal. The cost of EBI is only $2.29 per animal.

In Fakkema's ideal world, all the nation's shelters would switch to EBI and would include as part of the protocol the use of a pre-euthanasia sedative. Of course, in Fakkema's ideal world, shelter workers would no longer face the daily task of killing healthy animals. Looking at the numbers of animals being killed a present, this ideal world seems to me an awfully long way off. But Fakkema offers a glimmer of hope: the data, he says, suggest that we are headed toward a day when healthy animals are not being routinely euthanized. The most recent data—the 3.4 million animals—represent a new low for shelter killing. Our county only recently started spaying and neutering in volume, and it takes about ten years to see this reflected in the number of surplus animals. San Francisco, he said, is very close to replacement rate (though Fresno, just down the road, has the highest euthanasia rate in the country). I hope he is right.

ASSESS-A-PET

While researching this book, I came across a documentary film called *Shelter Dogs*. It takes place at the Rondout Valley Kennels in upstate New York, and the central figure in the film is the kennel owner and director, Sue Sternberg. Sternberg and her staff make efforts to adopt out their dogs, but when these efforts fail, they proceed, without looking back, toward killing.

The footage in the documentary is heartbreaking. The camera leads us past a row of concrete runs in a nearby no-kill shelter, where some dogs, we are told, will spend their entire lives. The quality of life for these dogs is very low. They are under a great deal of stress, have essentially no social interaction with humans or other dogs, and have very little in life that offers them pleasure. Sternberg's uncompromising moral stance—that ending the lives of unadoptable dogs is more humane than leaving them in the shelter environment—has a certain pull.

Shelter Dogs left me feeling morally unsettled for a variety of reasons. But one aspect of life and death at Rondout Kennels left me downright uncomfortable: the rigid adherence to a program of temperament testing (which I mentioned, briefly, in the context of animal personality in chap. 4). Temperament testing, or TT as it is called in shelter lingo, attempts to measure different aspects of a dog's personality, such as shyness, aggression, reactivity, and protectiveness. (It is worth noting some confusion and inconsistency in use of the terms "temperament" and "personality." Temperament is used more frequently in shelter settings, perhaps because personality has an overly human ring.) A test might, for instance, expose a dog to various kinds of stimuli or threats—strange or loud noises, visual surprises such as an umbrella opening suddenly, the approach of a weirdly dressed stranger. A rubber hand attached to a broomstick might reach out and try to touch the dog, to see if the dog bites. Many shelters use some kind of temperament testing to assess adoptability, trying to determine whether the dog is dangerous or is likely to have serious behavioral problems. The tests are relatively quick and easy—maybe fifteen to thirty minutes, tops.

Sue Sternberg claims to be a nationally recognized expert in temperament testing, and she peddles her trademarked Assess-a-Pet program (yes, trademarked!) through workshops held around the nation. Her website tells us, "Sue's nationally known temperament test offers shelters a set of procedures and responses in which to understand the behavior and future success for each dog in a home. Assess-a-Pet™ identifies congenial, family pets in animal shelters so that animal shelters will become the best places anyone can find their new dog."

I'm not an animal behaviorist, so I can't gauge the reliability or practicality of the Assess-a-Pet system. And although I see the value of temperament testing—shelters don't want to adopt out animals that will maul innocent children—there is also something about this that makes me uncomfortable. It seems too prepackaged, like the perfectly formed pastry from a vending machine. Even more, I find the emphasis on congenial, malleable dogs disturbing. It seems an affront to the diversity of dog personalities to expect all dogs to fit a certain (friendly) profile. There aren't that many naturally malleable and congenial humans; why expect so much more of our dogs? And seriously: if some stranger tried to reach out and pet me with a big rubber hand attached to a broomstick, I would be likely to bite, too.

These trait assessments can seem arbitrary—a stressed-out dog is given a thirty-minute test in a highly artificial setting—especially because the results of such tests can determine whether an animal lives or dies. Although many shelters swear by their methods, the validity of these temperament tests has very little empirical grounding. In other words, no one really knows whether or not the "aggressive" dogs who are euthanized would ever have bitten anyone. Still, many animal advocates and scientists believe—correctly, I think—that continued research into animal temperament can have a beneficial effect on animal welfare. Most obviously, temperament testing needs better empirical grounding: we need to understand whether and which measures of temperament actually map on to future problematic behaviors. Even more, understanding that animals are unique and full of personality can help foster a better appreciation and sense of empathy toward them. It can also be a useful tool for matching canine and human personalities in ways that will maximize the happiness of both. (On temperament testing, see Charles Siebert's *New York Times Magazine* essay, "New Tricks.") More to the point for me, the better we understand our animals, the better situated we will be to make decisions at the ends of their lives.

TO KILL OR NOT TO KILL, THAT IS THE QUESTION

Target is a hero and looks the part: golden-haired and with the distinguished lines of shepherd in her face. She is one of three strays from Afghanistan who thwarted the attempt of a suicide bomber to blow himself up inside an American military barracks. As the man tried to enter the compound, the dogs stopped him short, snarling and barking in the doorway. He detonated himself at the entrance and killed only himself and one of the three dogs. Target and Rufus, the survivors, were adopted by soldiers and eventually flown back to the United States, where they became minor celebrities. Target even appeared on Oprah.

After escaping from the yard of her new home in Florence, Arizona, Target was picked up by animal control and taken to the local shelter. She had no collar and no microchip, so her photo was posted on the shelter's website in hopes that she would be claimed by her owner. This was on Friday. On Monday, when her owner went to the shelter to look for her, he found that she was PTS—shelter lingo for "put to sleep." The shelter worker in charge of euthanizing animals that day had apparently picked the wrong dog out of the pen. The tragic story of Target has galvanized what has come to be known as the No Kill movement.

There are those, like Sue Sternberg from *Shelter Dogs*, who believe that killing in shelters is the lesser of two evils. People for the Ethical Treatment of Animals is also pro-euthanasia: they would rather see a dog euthanized than live her entire life in a kennel or loose on the street where she might "starve, freeze, get hit by a car, or die of disease" or, if not these, then the even more unpalatable, "be tormented and possibly killed by cruel juveniles or picked up by dealers who obtain animals to sell to laboratories." People for the Ethical Treatment of Animals concludes that, "because of the high number of unwanted companion animals and the lack of good homes, sometimes the most humane thing that a shelter worker can do is give an animal a peaceful release from a world in which dogs and cats are often considered 'surplus.'"

The shelter industry is organized around the following truth: pet overpopulation is a huge problem and the only viable solution is to kill unwanted animals. Until people become responsible about animals (and let's not hold our breath), we are stuck with the unfortunate task of killing massive numbers of Unwanteds. Incidentally, it isn't only cats and dogs who die in shelters: many ferrets, guinea pigs, rats, mice, hamsters, birds, and other critters wind up in the shelter's night drop box or at the relinquish desk.

This entrenched pattern of thinking—and killing—has been challenged over the past decade by a new philosophy of sheltering called the No Kill movement. Spearheaded by attorney and activist Nathan Winograd, No Kill offers a different way of thinking about animals in shelters: first we debunk the myth of overpopulation and then we rethink how shelters are run, day to day (fund-raising, business models, community interactions). According to Winograd, 90 percent of all shelter animals are "savable" (i.e., they are not hopelessly ill, injured, or vicious). If we could increase the market for shelter pets by a mere 3 percent (by, e.g., reducing the appeal of buying from pet stores and breeders), we could eliminate shelter killing. The problem isn't animal overpopulation, per se; it's mismanagement of shelters, alongside an overabundance of breeders peddling animals to pet stores and individual buyers.

It seems to me that Winograd is well worth listening to. Nobody likes the fact that animals are killed in shelters, but the widespread assumption that such killing is "sad but necessary" may amount to a monumental cop out. Rather than wringing our hands over the necessity of killing these animals, our energy might be better spent trying to formulate solutions.

COLLECTIVE GUILT

In September of 2010, the manager of the Miami-Dade animal shelter in Florida euthanized a four-month-old puppy on live television (the segment has since been withdrawn). The purpose of the spectacle, according to shelter manager Xiomara Mordcovich, was to show people in the community what was happening inside the shelter, to prick the collective conscience about pet overpopulation, and to encourage responsible pet ownership. "People need to see what happens here, and they need to understand that this is the consequence of what happens in the community out there. This is what we do to our best friend."

I found the news clip unsettling, as did a great number of viewers who wrote in to the station. Why? Most obviously, it is disturbing to watch an animal die. (I find it much easier to watch people in movies die than animals.) But worse than this, I think, is the sense that this particular puppy was made an example of and that his own undeserved death became a public spectacle. Nevertheless, from this gruesome event we gain a keen sense of this shelter's level of desperation and should appreciate the desire for transparency and the call for greater accountability toward animals.

HUMAN COSTS OF ANIMAL EUTHANASIA

To those who work in professions that involve caring for animals—primarily veterinarians and shelter workers—falls the grim task of ending these animals' lives. This paradox has even been given a name by sociologists: the caring-killing paradox. Even for veterinarians such as Kathy Cooney, who sees her work as offering an important gift to animals, her daily job carries personal costs, including a sense of compassion fatigue. Even though she receives a great deal of positive feedback from her clients—she told me that vets receive more thank you cards for euthanasia than for any other procedure—her work is hard.

For shelter workers, the personal costs are particularly high, since they are asked to perform such a large number of killings, often on healthy, adoptable animals. Although they may rationalize that death is in the best interests of the animal, this cannot remove the anguish of their task. Sociological research on shelter workers suggests a high level of work-related stress, unhappiness, and moral discomfort. The AVMA *Guidelines on Euthanasia* even warn that "constant exposure to, or participation in, euthanasia procedures can cause a psychologic state characterized by a strong sense of work dissatisfaction or alienation which may be expressed by absenteeism, belligerence, or careless and callous handling of animals." A study out of North Carolina found that shelter workers are at risk for high blood pressure, ulcers, depression, unresolved grief, substance abuse, and suicide.

On a related note, research in the United Kingdom found that veterinarians have high suicide rates—four times higher than the general population and twice as high as other health professionals. Drug overdose is the most common method of suicide, presumably because vets have lethal drugs readily available for animal euthanasia and have a good understanding of how to use the drugs effectively. On the human costs of euthanasia, Krista Schultz writes in a veterinary newsmagazine,

> Veterinary surgeons may experience uncomfortable tension between their desire to preserve life and their inability to treat a case effectively, which may be ameliorated by adapting their attitudes to preserving life to perceive euthanasia as a positive outcome. This altered attitude to death may then facilitate self-justification and lower inhibitions toward suicide as a rational solution to their own problems.

HUMAN AND ANIMAL: DARE TO COMPARE?

I have been surprised by this: people are eager to talk about the choices they have made at the end of their animal's life. Since I started working on this book, I have taken every opportunity I could to ask people about their experiences and hear their stories. Sometimes I ask, but more often people say without provocation something like this: "I just wish we could be so compassionate with people." Many have had the experience of watching a (human) loved one die a protracted and ugly death. Almost everyone I talked to—above all the veterinarians—spoke in favor of assisted dying for humans. "There should be a way out," people say.

Tom Watkins, reporting for CNN on the British study on veterinary suicide, remarks: "Euthanasia is a frequent duty of veterinarians, and the action must often be explained, encouraged and justified to clients. This constant interaction, performance and support of euthanasia in the animal population may affect profession attitudes on death in general. A small-scale European study determined that 93 percent of veterinary health-care workers interviewed approved of human euthanasia." I'm intrigued because this seems to confirm a refrain I heard over and over as I researched this book: If it is compassionate for our pets, why isn't it compassionate for our human fellows?

My sample is certainly skewed, and what I say here is pure speculation. But it makes me wonder: Does being a pet owner—having perhaps made the decision to euthanize an ailing or dying animal—make one more open to euthanasia in general? Does being a vet have a similar effect on one's attitude toward end-of-life care for humans?

It is interesting to observe that physicians for humans—at least when they are surveyed by academicians—offer quite a different picture. A recent survey of US physicians found that 69 percent object to physician-assisted suicide or PAS (which is close as we come to euthanasia), and fully 18 percent object to terminal sedation and 5 percent to withdrawal of life support. The primary arguments given against PAS are these: pain medication is good enough that there is no reason for a patient to be in intractable pain (thus there is no reason that they should desire to die); physicians might incorrectly diagnose terminal illness; PAS violates the role of physician as healer; and finally, we have what is known in bioethics as a "camel's nose under the tent" argument, from the old Arabian proverb "If the camel once gets his nose in the tent, his body will soon follow": if we allow PAS for some patients, this will lead ineluctably to the killing of patients who do not want to die.

Jerrold Tannenbaum observes that discussions of euthanasia by medical ethicists fail to mention veterinary medicine or the euthanasia of veterinary patients. "This apparent lack of interest is startling," he writes, "because many objections to euthanasia in human medicine stem from the fact that human medicine has had little experience with it." He goes on: "However, there is a healing profession with extensive experience relating to the euthanasia of its patients. These doctors have long had to worry about when (if at all) euthanasia is justified, how to perform it, and what effects it can have on those close to the patient." Some cross-disciplinary discussion would be enlightening for both sides.

Tannenbaum's reflections on veterinary euthanasia paint a mixed picture. The experience of veterinary medicine shows that a profession allowed by law, its own official ethical standards, and societal attitudes to kill its patients may well kill too many. Fears in human medicine about who might be responsible for overutilization of euthanasia could be misplaced: it is the clients, not the vets, who ask for euthanasia. Practitioner-induced euthanasia is not inevitably associated with disrespect and devaluation of the patient. There is a link between the value people place on a being (or kind of being) and their willingness to choose euthanasia for it. And finally, money is a significant motivation for euthanasia in veterinary practice: people often choose to euthanize an animal rather than pay for curative or palliative treatment.

While we may want at least to consider being more open toward human euthanasia, we may conversely want to show more restraint when it comes to our animals. We may want to hold back on that blue needle a little more often. There are things we have learned from our care of humans that might inform how we approach end of life with our animals. Research has found that it is people who are thriving who most strongly advocate for the legal option of assisted death; once a person becomes seriously ill, their perspective often changes, and they put up with far more than they would have predicted. Yet with our animals, we assume that their level of tolerance for pain and suffering is much smaller than ours and that death will be a welcome relief. We think about their dying in the same way a healthy able-bodied person thinks about her own assisted death, not like a dying person, whose perspective on the process is certainly going to be different and, of course, in many ways more authentic. And we don't, of course, think about it like a dying animal—whose mysteries we can never fully unravel.

The Ody Journal

NOVEMBER 29, 2010

Woke up about 12:45 last night and went to check on Ody, and I'm glad I did. I'm not sure why I woke up or why I got out of bed because all was quiet. As always, the smell hit me first. Ody had gotten trapped again between the wall and a leg of the piano. Poop was smeared all over the base of the lamp, all over the cords, all over the floor and wall, and all over a large corner of the carpet under the piano. Ody looked utterly pathetic as I untangled him and helped him out. First I let him outside, to go to the bathroom with dignity. Then he had a long bath. And then I stood, for a long time, just looking at the mess, not sure how to clean it or where to start. Finally, I donned some gloves and got some rags and started with the cords, one by one. Ody must have been there for a while because the poop was partly dried and very hard to clean off.

About forty-five minutes later Chris woke up and came in. He helped lift the piano legs one by one and we pulled the carpet out. It has to go in the trash—there is just no way to clean that much smeared, ground-in poop.

We lifted Ody onto the couch in the family room and he immediately settled down and fell asleep. After showering and changing into new pajamas, I took a blanket and pillow out to the couch and settled down next to Ody. I wanted to make sure he didn't get up again and need help. And I didn't want him to be alone.

Chris thinks it is time to call the euthanasia vet. I have to say that after yesterday and last night, I'm starting to come around, though there is part of me that still resists. I was going to say, "I'm not sure I'm ready," but I realize, as the words form, that it isn't me who needs to be ready, it's Ody. My readiness should have nothing to do with it.

I was thinking, as I lay on the couch next to Ody: this reminds me of the birth of my daughter. I really wanted to have a "natural" childbirth, with no drugs, no fancy stuff. I wanted to have the full human experience. I was prepared, and I was tough. Until I started having contractions, and after about eight hours when the doctor recommended an epidural to get things moving, I said "to hell with natural." I've wanted Ody's dying process to be "natural," too. I have this image of him gradually slowing down, eventually not getting up (lying peacefully on his oatmeal dog-bone bed), and then, after the appropriate amount of time (three days?) gently passing on. But the reality of the process simply may not match my ideal. *My* ideal, I must remind myself . . . not Ody's.

I called Home to Heaven, after several hours of delay. All morning Ody has been pacing and panting and falling over. I described to Kathy Cooney what was happening with Ody. She didn't advise me one way or another, but she said some things that made me feel that euthanasia will be better for Ody than natural death. It is very stressful for him that his body is not doing what it should (she could hear him panting in the background). Because his heart is still strong, his decline will be long and hard. He'll become less and less mobile and will start having more pain; he may get sores from being bed-bound; if he remains this anxious and worked up, he may start having seizures.

She said someone could come out today. I said no, let's wait until tomorrow. I need time to say goodbye. But I called Chris and he said that it would be kinder to Ody to do it sooner. This is for him, not us, he reminded me. Chris was sobbing, which somehow reassured me that this was the right decision.

I called Kathy Cooney back and said we would do it tonight instead. Kathy wasn't available, but she said that one of her other vets—Dr. Michaelle—would come between 6:30 and 7:00. After working out details of where and when, she told me that they would make a paw-print impression of Ody's foot, and she asked what I wanted to do with the body— cremation, I said—and did I want the ashes? She explained that the euthanasia will cost $200 and it will be $100 more for private cremation. "Fine, fine," I murmur.

And all of a sudden it hits me, like a ton a bricks. *We're really going to do this.*

All day long Maya has been standing right next to me—totally under-
foot in the kitchen and as I sit at my desk, just standing there to my right,
staring straight at me and lifting her paw occasionally to scratch at my
leg. Does she sense that something is not right with Ody? What does she
know?

It has been a very strange afternoon, anticipating what will happen this
evening. I've felt slightly sick all day. After his difficult morning, Ody finally let me put him on the couch and he settled into a deep sleep. I've
called the few people who would want to know in advance—my parents,
Liz and Craig, my brother. Others I will tell over time, maybe with some
kind of written note. I spent a little time playing "Old Blue" on the piano
for Ody.

I told Sage when I picked her up from school. She seemed surprised,
more than anything. "Really?" was her response. And then, in tweenager
fashion: "Well, this is really going to ruin my day!" She was very sweet to
Ody when we got home and pulled his Christmas present from under the
tree and opened it for him. It was some nasty gravy-based Fancy Feast—the
best of canine junk food. He lay on the couch and licked it off a big spoon.
Then she curled up next to him on the couch, with Maya and Topaz all in
the row, and we read *Three Stories You Can Read to Your Dog*. One is about
scaring away an intruder who knocks on the door, one is about a magic
tree that grows bones, and one is about becoming a Wild Dog for a day
and then going back home for dinner. The final lines of the book seemed
strangely appropriate:

> You were all tired out.
> Being a Wild Dog is the hardest work of all.
> You curled up in your bed.
> And you went to sleep for a long, long time.
> The End

Ody seemed so calm and peaceful and warm in his sleep—not a dog
needing to die.

But as if to remind me why the vet is coming, Ody fell over when he
finally got up off the couch. He lay sprawled out on the office floor, kicking
his legs in the air, until I could get to him and lift him upright.

He enjoyed a nice last dinner—hamburger and rice and some salami
slices and a piece of cheese. But his breathing sounds quick—as if he's

scared. Now that he's up again, he can't settle back down. The time is drawing close, and the pit in my stomach gets heavier.

How do I even begin to write this? I've been avoiding sitting down at the
keyboard because it feels so raw, but I'm afraid to lose the details of Ody's
final night.

The afternoon was so strange yesterday, inching along toward 6:30. I
found myself looking at the clock over and over, each time with a deep-
ening sense of apprehension. Ody was up and about when people began
arriving. My parents came first. Then Chris came, about 6:25. And right
at 6:30, the vet. I saw her headlights out front and had the door open for
her. She was a young woman, light-colored hair pulled back in a tight clip.
There was a good deal of commotion when she walked in—Topaz and
Maya barking, people milling about. I have such a vivid picture of Ody at
this point: he was standing in his halfway squat by the piano, just watching.
Chris started talking to Dr. Michaelle, and I overheard something about
where Ody's ashes would be delivered. But my attention now was on Ody.
He lurched through the kitchen toward the back door, intent on going
outside. I opened the door and followed him out. It was bitterly cold. Ody
limped onto the patio and crouched there, looking at me. I walked over
and sat on the cold stone next to him and buried my face in his neck. I
wanted to stop time, right there with Ody. We were safe in our own world
for that moment.

Chris stuck his head out and said, "The vet's ready to start." Ody almost
seemed to know what was about to happen because he didn't want to
come back inside. Always he will follow me back into the house, but not
now. Finally I had to step behind him and guide him toward the door. I was
escorting him to his death, and I didn't like the feeling of it. We decided
to do the euthanasia on the couch in the office—one of Ody's favorite and
most comfortable places to sleep. Chris carried Ody in and laid him down
on an old purple blanket. Chris knelt on the floor by Ody's head and I sat
next to Ody's tail. Sage perched on the arm of the couch by Chris, holding
Henry, her hairless rat, for emotional support. We locked Topaz in Sage's
room, thinking that he would be disturbed by the vet handling Ody. But
Maya was there. I called her over and she climbed up on my lap and curled
into a little ball, her head resting on Ody's back. Ody fell asleep almost the
minute he lay down, despite all the commotion in the room.

The vet knelt down by us and explained what she would do. First she would give Ody a sedative that would make him relaxed and unaware of what was happening. After about five minutes, when the sedative had taken full effect, she would place a catheter in Ody's leg and give him the final injection. Chris asked how long it would take for Ody to die. "It's pretty instantaneous," she said.

She asked if we were ready, and through tears we all nodded. She placed a small needle in the skin over Ody's ribs and slowly pushed in the sedative. Ody seemed still to be sleeping when she gave him the shot and didn't react. After some long moments, his breathing became heavier and the muscles along his ribs twitched. Dr. Michaelle said that this was a normal response to the sedative. Gradually, Ody's sleep deepened. After four or five minutes, the vet took out some clippers and said she would make sure the sedation was complete. The clippers began to buzz. Maya lifted her head, but Ody slept on. The vet shaved a small patch of fur from Ody's hind leg—again, he showed no reaction. She dropped the red fur into a small plastic tub. She said he was ready, whenever we were. All this time we were stroking Ody and telling him goodbye.

She put a rubber tourniquet above the shaved patch and tightened it. After feeling around for the vein, she placed the catheter, in that thin place on Ody's Achilles. She took a syringe with clear liquid, and Sage asked, "What's that?" The vet explained that she was going to flush the catheter. "It's just like water," she told Sage. "I'm making sure the medicine is going to go in just right." Then it was time. She lifted the second syringe—the blue one—and stuck the needle into the catheter and gently pushed the blue into Ody's body. Ody took three or four more slow breaths, gave one sharp inhalation, and then grew still. It was very fast—maybe twenty seconds. Maya lifted her head just after Ody stopped breathing, and cocked it to the side, as if aware of some change.

It was odd, but Ody had indeed changed. Even though he was in the same position, his jaw was now slack and his eyes had gone blank. His eyes remained open but seemed to have sunk down into their sockets. I pulled his catfish down so it would cover the brown stub of his tooth.

I stayed where I was, not wanting to move, not wanting to enter a new phase in which Ody would no longer be there. The vet went into the other room so we could have some time with Ody and, she said, so she could prepare the clay to make his keepsake paw print. Chris looked up at me and said, "Do you realize that Ody never learned to scratch himself without falling over?" I laughed a little through my tears. It's true. He never learned.

I asked Chris to let Topaz come in and see Ody—I wanted to give Maya and Topaz both a chance to understand that Ody had died, if they can understand such a thing. I didn't want simply to have Ody disappear. Topaz's reaction was interesting: he ran in, ears back, looking very upset. He barked once or twice—perhaps just warning us that a stranger was still in the house—and ran around the room, looking concerned. He sniffed Ody's face, and then sniffed his leg right where the catheter had been, and then he ran under my desk and lay down, ears still laid back. Chris took him back to Sage's room so the vet could finish up. Maya stayed on my lap, her head still resting on Ody's back.

The clay paw print was a nice gesture but a little strange. The vet had to push Ody's paw down hard to get an impression in the thick clay.

Chris got up and said, "I'll carry his body out to the car." We wrapped Ody in the purple blanket and Chris lifted him. His head hung down, tongue out, and he really did look dead. I found it hard to watch. Chris took Ody out and put him in the back of the vet's car. I hugged Ody one last long minute and told him, one last time, "You're my best boy." He was still warm and soft. I know it is just a body, just skin and bones, but it broke my heart to think of Ody stiff and cold, being lugged by some stranger into the crematory. This wasn't how I had pictured the very end.

We all went about our grieving in different ways. Sage went into her room and played a computer game. Chris and I took the other dogs for a walk and then he cooked some dinner and turned on the television. I retreated to the bedroom, put on my pajamas, and crawled under the covers and cried.

DECEMBER 1, 2010

The next morning Chris told me there were some booklets on the table, left for us by the vet. About grieving. I'm not ready to look at them yet. And I'm not ready to bake Ody's clay paw print, to make it permanent.

It feels really quiet. I actually missed getting up in the middle of the night to let Ody out, and I even miss the panting. I feel such a mix of emotions: sadness and also a sense of relief, both for Ody and for myself. Ody has been one of my greatest loves and also my millstone, for fourteen long years. I know Ody was suffering, and I feel that we made a good decision. I'm also disappointed that we resorted to euthanasia.

Maya has been afraid to come into the office today. She sits just down the hall and whines, and when I call her in, she just makes a funny little sound. She lies in the hall with her head on her paws, staring at me and crying. Does she remember that Ody died here? Maya and Topaz both seem a bit subdued. I don't think Topaz is sad about Ody—I think he just follows Maya's lead. I've noticed already some jockeying for position and shifting of the balance of power. Even though Ody was at the absolute bottom of the hierarchy, his absence still seems to have shaken things up a bit. Topaz, it seems, is going to be the new boss.

Mostly I think we made the right decision, but sometimes this feeling washes over me all of a sudden and I think, *Oh, God. What have we done?* Couldn't we have helped Ody live on a little longer? I was rereading the story of Baxter, the hospice hound, this morning, and it made me have a crisis of confidence. Baxter was nineteen when he was filmed doing his rounds at the San Diego hospice, and he was so arthritic that he couldn't walk. His owner had to pull him through the hospital in a wagon. Couldn't we have gotten a wagon for Ody?

I went through all our old pictures and pulled out all the ones I could find of Ody. It amazes me to see him in his prime—so different in physical body, but still the same look in his eyes: angst. Even in his happiest pictures, there is a shadow over his soul. Although I now have pictures ready, I haven't been able to work on Ody's memorial card. I'm not quite ready to face this.

When Sage got home from school, she went straight to her room and stayed in there for several hours. Turns out she was making a movie about Ody on her computer. Her rats have movies, too. This is her way of remembering.

I think about how the vet shaved a patch of hair from Ody's leg, where she inserted the catheter. She dropped the hair in the plastic tub. I so wish I had that fur now . . . the last of the Ody seeds are gone. And I have none.

DECEMBER 2, 2010

Sage wrote a eulogy for Ody last night. Each sentence is written in a different color pencil.

Sniff the humid
air in Chalco Park,
Explore the wild

meadows of Estes Park,
You are Ody experiencing
the wonders of Life.
Ody: a wonderful dog.
Regal stance and
flowing red fur.
He was our Ody, our
Odysseus. Our toothless
wonder.
In his youth he was
energetic and always friendly.
Even through old age
he was spirited.
Ody was determined
to eat, escape, and destroy.
However, our toothless
wonder always came
back one way or another.
We love our Ody, our
Odysseus, our toothless
wonder.
I—we—will miss our
Ody, our Odysseus, our
toothless wonder.
Ody was truly
special, we will
never meet a dog
quite so mischievous
and persistent.
And he was so
loveable, so sweet,
but yet so naughty.
Ody, Ody, Ody we will
miss our Ody.
Goodbye Ody, Odysseus,
our toothless wonder.
We will miss you.
We love you.

A friend gave us Cynthia Rylant's book *Dog Heaven* as a sympathy gift. My favorite page is the one about clouds. One of the dogs in the painting reminds me of Ody, and one looks just like Topaz and one like Maya. "God turns the clouds inside out to make fluffy beds for the dogs in Dog Heaven, and when they are tired from running and barking and eating ham-sandwich biscuits, the dogs each find a cloud bed for sleeping. They turn around and around in the cloud . . . until it feels just right, and then they curl up and they sleep."

DECEMBER 7, 2010

I went to pick up Ody's ashes yesterday at the animal hospital. He was cremated at a place called Pennylane. The ashes came back in a pretty, cylindrical wooden box, tied with a yellow ribbon. Along with the ashes, we have a certificate of cremation.

> Date: December 4, 2010
> This is to certify that your faithful companion
> *Ody*
> was this day cremated in our crematory, with respect, care
> and observance of all legal requirements.
> Date of death November 29, 2010

It's been a week and a day now. Do I think Ody had a good death? Yes. Am I sure I did the right thing at the right time? No. I'm finding that I have moments of doubt, anguish even, over my choice. And these seem to come more frequently than in the day or two after his death, when perhaps the overwhelming feeling was a sense of relief. Walking out of the vet's office with Ody's remains, I felt heavy with sadness. I felt, in that moment, that I had let Ody down. That I could have fought harder for him. But to what purpose, I'm not sure. Maybe the fighting would have been for me more than for him.

When I tell people about Ody, I try to avoid euphemism. I try not to say, "We had to put Ody down." But this is what often escapes my mouth. It sounds so jarring: "I euthanized Ody. I chose to end his life." But there it is.

7

Remains

The book is called *Animal Folk Songs for Children*. Half of the red cover is missing, and the remaining half is faded and scuffed. Some pages are torn, and some have been decorated in purple crayon by a younger me. It is full of folk songs about animals: "Mister Rabbit," "Cross-Eyed Gopher," and "Let's go a Huntin'." I love many of these songs, but there is one, in particular, that I love beyond all others. It is a traditional Mississippi folksong called "Old Blue" and singing it makes me feel like a dog howling at the moon. I grew up listening to a Burl Ives recording of this song. I would ask my mother to play "Old Blue" on the piano and I would sing along. "One more time," I would say, over and over. The plaintive melody would run through my head at night as I fell asleep. The song is immeasurably sad, and it touched some deep down place in my soul.

The song is about a hound dog named Blue, who gets sick, is told by the vet "Blue, your huntin' days are done," runs all around the yard digging little holes, and dies. The final three stanzas, which take place after Blue's death, are my favorite.

Laid him out in a shady place,
Covered him over with a possum's face,
   Oh Blue, Blue, Blue, oh, Blue.
Dug his grave with a silver spade,
Laid him down with a golden string,
   Oh Blue, Blue, Blue, oh, Blue.
When I get to Heaven I think what I'll do,
I'll take my horn and blow for Blue,
   Oh Blue, Blue, Blue, oh, Blue.

I still have *Animal Folk Songs*. Every now and again I take the book off the shelf and play "Old Blue" on the piano and sing along. And invariably it makes me sad, especially the part about blowing my horn when I get to heaven and Blue will come running—which always leads to a quizzical look from my daughter, if she happens to notice the tears running down my face. "Mom," she'll say, "you're really weird." And perhaps I am odd to be so moved by song about a dead dog. But those who have experienced this loss first hand will know what I mean.

Just as dying is a biological process that occurs over time, death really doesn't end when the biological organism shuts down. I can identify the moments of Ody's biological death—the span of twenty or thirty seconds when his heart slowly stopped beating—and there was an awful sense of finality to those last breaths. But his dying seems to still be happening, in some peculiar way.

AFTERCARE

Respect for an animal doesn't end when the physical form dies but extends into how we treat the body. Children seem to have an innate sense of this. When Bubbles the goldfish dies, we must not simply toss the body into the kitchen garbage disposal or flush it down the toilet. Instead, we carefully place the body inside a tiny jewelry box, dig a hole in yard, paint a special rock to place over the grave, and say solemn words about what a remarkable goldfish Bubbles was and how much happiness he added to our lives. Respect can take many forms. For example, it can be as uncomplicated as Jerrold Tannenbaum's instruction to his students in *Veterinary Ethics*: we must close the eyes, put the tongue in the mouth, clean the body, and wrap it in a clean blanket. Or we can choose a more elaborate ritual, as in "Old Blue": we can cover our dog with a possum's face (which is just too bad for the possum) and lower him into the ground with silver strings.

Aftercare refers to all kinds of decisions one might face after a pet has died. How long will you stay with the body? Where will the body be kept? With what will you wrap or cover it? Will you have a funeral or memorial service? Coleen Ellis is arguably the nation's expert in animal aftercare and has taken it as her mission to help alter the landscape of death care for animals. Eight years ago, Ellis opened the nation's first pet funeral home, the Pet Angel Memorial Center in Indianapolis.

Based on her own experiences following the death of her terrier-schnauzer Mico, combined with her experience working in the human funeral business, she realized that the choices available after an animal dies are nothing like those offered for people. Animal bodies are "disposed of," not honored. But many "pet parents," as she calls pet owners, want to see their animal's body treated with respect, not dumped in the landfill. People don't know how to proceed after a pet has died, and vets aren't really equipped to help them. But who is? Ellis set out to fill this gap.

Ellis told me that some people feel uncomfortable using the same kinds of rituals to honor animals that we use to honor people. Funerals and wakes for animals are often furtive affairs, held in the backyard when the neighbors aren't looking (do they really want animal skeletons so close by?). We may be afraid to show others how deep our grief for an animal can run. But, she says, we shouldn't scoff at the application of death rituals to animals. We need permission to ritualize, to make the death of an animal companion meaningful, to honor them and our bond with them in ways the make sense to us. The ritual can be simple or complex, short or long—it really doesn't matter, as long as it shows our proper regard for the deceased.

She listed for me the most important services she provides for her clients at the pet funeral home. If the client chooses to have a funeral service, Ellis can arrange for a visitation period. The animal will be laid out in a coffin, and family and friends can come and make their final farewells. "People don't realize how important this is, until they do it," she told me. It can be particularly helpful for people who have had to euthanize their animal at a veterinary office, perhaps unexpectedly. They may have been too overwhelmed during the euthanasia to say a proper goodbye. It is particularly important for children, she says. Children need more than "Fluffy disappeared while you were at school."

Before the visitation, Ellis will prepare the animal by closing the eyes, putting the tongue back in the mouth, cleaning the face and body, putting charcoal down the throat, packing orifices that might bleed out, pulling in the legs, and generally making sure that the animal looks natural and peaceful. The animals is wrapped in a cozy fleece blanket and placed in a casket.

After the visitation, some families choose to hold a funeral or memorial service, most often at their home, in their backyard, or at a park. Ellis will help pet owners choose readings or prayers or eulogies and can suggest various nice touches such as candles and flowers and animal-themed foods (hotdogs, pigs in a blanket, and so forth). We've had many animal funerals in our backyard, but none as nice as the ones that Ellis describes to me. Our most elaborate, by far, was the funeral for Sage's first rat, Fuzzies, which took place in our backyard and was officiated by a Mormon missionary (who, it must be said, was quite uncomfortable with our request that he say a prayer over Fuzzies' body).

In addition to helping plan aftercare services for more than six thousand families, Ellis has helped mobilize a new arm of the pet industry. The Pet Funeral Home Directory now lists at least eighty pet funeral homes, and the number is growing yearly. Ellis's work has been so successful that she has moved away from running her own funeral home and now consults with others who want to start a pet funeral business. She also works with various veterinary colleges and clinics, educating vets in death care and bereavement.

A few years ago, Ellis founded the Pet Loss Professionals Alliance, an umbrella group for all manner of animal aftercare, working in partnership with the human arm of the aftercare industry, the International Cemetery, Cremation, and Funeral Association. One of the main tasks of the alliance is to give more formality—and more standards of ethics—to the pet-loss business. For example, the organization wants to see industry-wide agreement on terminology such as "individual cremation." Ellis would like to see each and every veterinary office have some kind of "senior package" with information about caring for an aging animal. This packet should always, she says, include a guide to planning for death and aftercare. Perhaps more than anything else, she wants to change the stereotype that pet owners who want a casket and a funeral service with nice flowers and candles are either eccentrics who see their pets as fuzzy children or lonely people who have no friends other than their dog or cat.

DISPOSITION OF THE BODY

The most important aftercare question is "what will happen to the body?" Even for those who are unsentimental about dead animals, there are reasons why disposition—or, to put it more crudely, disposal—of the carcass matters. (The term "corpse" is usually applied to human bodies, while a dead animal body is more often referred to as a "carcass.") As the Colorado State University Cooperative Extension explains, "Animal deaths must be handled properly for at least three important reasons": health (to limit the spread of disease), environmental protection ("nutrients as well as harmful materials released as dead animals decay can drain or be carried into nearby water"), and appearance ("people may find the sight of dead animals 'very disagreeable'"). They list the following acceptable ways of managing animal deaths: rendering, composting, sanitary landfills, burial, and incineration.

I'll talk in more detail about burying and burning, the two methods most commonly chosen for disposing of companion animal bodies. And I'll take a look at the world of rendering, which is a manner of "recycling" bodies that pushes the limits of respectful treatment. But first, let's take a detour through some techniques for preserving the body, if you simply cannot let go, or if you want a permanent and lifelike reminder of your animal.

LOVING AND LASTING

Taxidermy and freeze-drying offer ways to preserve an animal's body in perpetuity. Traditional taxidermy used to be the method of choice for people wanting a permanent keepsake. Taxidermy produces a three-dimensional replica of an animal. Sometimes the actual skin of the animal is mounted on some kind of frame; sometimes the animal is reproduced using nothing but manmade materials. Not surprisingly, the preserved animals tend to look stuffed.

A more lifelike option has become available that appears to be replacing taxidermy as the preferred preservation method: freeze-drying. The Perpetual Pet website boasts:

> Through the use of new techniques in freeze dry technology, we can offer a "Loving and Lasting" alternative to burial, cremation or traditional taxidermy. Freeze-dry pet preservation creates a lasting memorial and more importantly, preserves your pet in a natural state thereafter, without any alteration in appearance. This allows pet owners to see, touch and hold their pets, and in a sense, "never have to let go."

To freeze-dry an animal, you place it in a sealed vacuum chamber at very low temperature. Over time, "frozen moisture is slowly converted into a gaseous state, and then extracted." By removing the moisture, you stop the process of decay. The drying takes considerable time. For a large dog like Ody, it might take up to six months. Perpetual Pet suggests that you choose a sleeping pose, for the most natural-looking results, but they will accommodate other requests. I could have Ody frozen into his basic slither-over-the-fence posture and keep him in the backyard, or maybe have him frozen into a permanent begging posture and stand him right next to the dining room table. The website only lists prices for animals weighing up to twenty pounds; anything heavier and you must call for a quote. Maybe this service appeals especially to people with small pets, like cats or Yorkies, which would be relatively easy to display on a table or mantelpiece. A seven-to-ten-pound pet will cost $695. I'm guessing Ody would cost well over a thousand.

I asked Kathy Cooney if she knew of clients who chose freeze-drying or taxidermy and she said no. She has, however, had people request that she cut off an animal's tail, or sometimes a toenail down to the stub, so that they can save a part of their animal's body. She has also had people ask for whiskers, hair, eyelashes, and a skull. Coleen Ellis said she has had only two or three families, out of six thousand over five years, that chose freeze-dying.

Ellis tried to remain neutral when I pressed her about freeze-drying. "Doesn't it seem a little creepy?" I asked. She replied that she tries to respect whatever choices a family makes. She did admit, though, that she had tried to steer some people away from freeze-drying, simply because she thought it wasn't ultimately what they needed. One woman, for example, wanted to freeze-dry the family dog because she was worried about how her son was going to handle the dog's death and thought that being able to touch and stroke the animal would reassure him. Ellis felt that freeze-drying was not a good choice for this family because, by the time the dog's body was frozen and returned to the family six months later, the boy would likely have processed his grief and might actually find the return of his dead dog disturbing.

Unlike taxidermy and freeze-drying, which are available only for animals, cryonics has developed primarily as a way to preserve the human body, and cryonic preservation of pets arose only as an afterthought. Although cryonics and freeze-drying both involve extreme cold, they differ in important respects, both technical and philosophical. The Cryonics Institute explains:

Cryonics is a technique designed to save lives and greatly extend lifespan. It involves cooling legally-dead people to liquid nitrogen temperature where physical decay essentially stops, in the hope that future technologically advanced scientific procedures will someday be able to revive them and restore them to youth and good health. A person held in such a state is said to be a "cryopreserved patient," because we do not regard the cryopreserved person as being really "dead."

Cryonics would be most effective if it could be accomplished while an animal was still alive since tissues and brain cells would not yet have begun to deteriorate. But unfortunately, this is still illegal, for pets and people alike. For now, instead, in order to have your pet cryopreserved, you must keep the body frozen from the moment of death and carefully pack the body in dry ice so it can be shipped to Michigan. Excluding the cost of membership in the Cryonics Institute, Ody would probably cost about $6,500 plus shipping and veterinary costs. They say: "If these prices seem excessive and you can be satisfied with the possibility of someday having a clone of your pet, you can save your pet's DNA with the Cryonics Institute for only $98." The Cryonics Institute reports that at the current time they have fifty-eight pets and thirty-one pet DNA samples in cryostasis.

If you cryogenically preserve your pet, you will not have the pleasure of its company in your house; the body must remain suspended in a tank of liquid nitrogen at the cryonics facility at a carefully controlled temperature of −196° Celsius. But you can live with the hope that someday you could have your actual pet again—or an exact clone of your pet—once science has progressed enough that we can reanimate frozen bodies or somehow upload animal souls into a brave new cyberuniverse.

RENDERING

Rend: To separate into parts with force or sudden violence; to tear asunder.
To render: to melt down; to extract by melting.

There is a reason I avoid eating anything made with lard, aside from the obvious fact that I follow a vegetarian diet. I don't want to eat anything that came from a rendering plant. My stomach is too weak. Rendering is the process of converting dead animals and animal by-products into usable proteins, with "usable" here being very loosely defined. Horses and other very large animals, which are cumbersome and expensive to bury or cremate, are often sent to the rendering plant—euphemistically known as the glue factory. (Have you ever noticed that the Elmer's Glue logo is a little picture of a bull?) And when a dog or cat dies at a shelter or a veterinary office, which may have large numbers of critter bodies to deal with, there is a chance that it will wind up at a rendering facility. Rendering is by far the most economical way to dispose of a body.

Most of the animal parts that end up in rendering facilities come from slaughterhouses, which collect and transfer all the unusable parts of the cows, pigs, sheep, chickens, turkeys, and so on, as well as the animals that are sick and cannot legally be processed into meat. Other sources of waste animals include expired meat from grocery stores, trimmings from butcher shops, dead animals from zoos, and—you guessed it—dead cats and dogs from veterinarian offices and shelters.

I once watched with fascinated horror an episode of the television show *Dirty Jobs*, where the host spent a day working in a rendering plant. The host helped unload a cow carcass from a truck and remove the skin by punching a hole in the hide and blowing it up with an air pump, just like a balloon, to rip the skin from the muscles and bones. After sawing off the now-separated hide, he helped haul the skinless carcass onto a conveyer belt and watched it ride up into the jaws of a giant woodchipper-like machine that ground it into fine bits. Then our host went into a different part of the factory, where the animal puree was put into an enormous caldron and cooked and cooked to release the fat and drive off the moisture. Once the fat was pressed out of the solid material, we had the "cracklings" that would then be ground up and used to make meat and bone meal. I actually lost my appetite for the rest of the day, and thinking about the show still scares me.

But a small voice in my head objects that this is simply recycling and that if we are going to use animals, it is really good to use everything we can. It is more respectful to the animal than simply throwing parts away. The meat and bone meal can be used in all manner of ways, including the manufacture of processed dog and cat food. Kathy Cooney stressed to me that we *need* rendering facilities, much as we may not like the idea or the smell of them. Our society has massive numbers of animal carcasses to deal with, and rendering facilities perform an important function in handling, safely, all these dead bodies. She told me that the number of rendering fa-cilities in the United States is on the decline and worries that this may be-come a problem.

BURIAL

Various animals throughout human history have been interred with great care, and intentional burial of dogs by humans dates back at least to the late Pleistocene. Archeological evidence suggests that dogs were even bur-ied together with humans, in the same cemetery. A team of archeologists studying two such burials in Cis-Baikal, Siberia, believes that as soon as formal cemeteries developed in this region, about seven thousand years ago, some canids began to receive mortuary treatments similar to those performed on humans. Some human graves in the area contain "canid re-mains modified into body ornaments or implements."

I always imagined that I would bury Ody high up in the hills, behind our cabin in the Rocky Mountains. This is just what you do: you take care of the body yourself. You bury your animal on your own land or in your backyard. Two of my childhood dogs—my little cocker spaniel Benny, and our shepherd/husky Nathan—are buried up at the cabin. For all the small critters we've had and lost—all Sage's rats, her gecko, her hermit crabs, her goldfish and guppies—we've simply dug a grave out behind our house along the fence.

I discovered during my research that burying animals in the backyard is not so straightforward. Not everyone has a backyard, and some backyards are too small to bury much of anything. Furthermore, in some counties, burying animals on your property is illegal. Zoning ordinances in Boulder County, where I live, forbid any animal burials (oops). But the more rural Larimer Country—where the cabin is located—does allow burial. Neigh-boring Adams and Weld Counties also allow burial, but there are certain regulations. For example you must be at least 150 feet down gradient from water sources and bury all parts at least twenty-four inches deep. Even if you feel fine about burying in your backyard and it happens to be legal, you might worry about then moving and leaving your buried animal be-hind (and what the new owners might think when they dig up the yard to re-landscape). There is also a real danger, when one goes digging around in the yard, of hitting buried utilities. Nor can you simply head out to the forest, since burial on public lands is nearly certain to be against the law. Furthermore, if an animal happens to die in the winter, as Ody did, and you live in a cold area, home burial may not be feasible. Finally, as one last consideration, the problem of secondary wildlife death from animal burial is quite serious. Euthanasia solutions can remain potent in an animal's body for as long as two years and can harm and even kill wild animals who dig up and eat contaminated carcasses. This is one reason vets who need to euthanize large livestock in the field will use a gun, rather than euthanize with drugs.

The more straightforward burial option is to find a pet cemetery. Our modern iteration of an animal burial ground is available in many areas. Over six hundred pet cemeteries are up and running in the United States, and Colorado has five that are accredited by the International Association of Pet Cemeteries and Crematories. The closest two are Precious Memories, in Fort Collins—about forty-five-minute drive—and Evergreen Memorial Park, about an hour's drive. Some human cemeteries have special pet sections, and in a few states, people and their pets can be buried together.

I researched what it would be like to have Ody buried in a cemetery, and I didn't make it past the first step—purchasing a casket. From the multitude of pet memorial products websites, I chose petsweloved.com. For a handcrafted, cherrywood-finish casket big enough for Ody (and for which I would choose the ivory lining), I would pay $334.95 plus shipping and handling. The Everlast, in polyurethane, is the cheapest option. Ody's large casket would cost $199.95. Not wanting to offend Ody's spirit by choosing the cheapest coffin, I would probably opt for the midlevel choice, the Furever made of High-Impact Styrene. I wondered, though: would Ody really want to spend the rest of eternity locked away in High-Impact Styrene? And things were looking expensive. The casket was just the beginning. The cost of burial at Evergreen would be $325 for perpetual care with flat-to-the-ground marker and $550 for perpetual care with upright marker. Granite and bronze markers start at $450, and we would have to pay an extra $35 for winter grave preparation.

Cemeteries are a good option for some people, especially the financially comfortable, but there are cautions. Pet cemeteries are not subject to the same regulations as human cemeteries. If a cemetery changes owners, or if the property is sold, the land may legally be used for other purposes. In some cases, people have been told, years after the death of an animal, that they will need to dig up their pet if they want to keep the remains. It is worth asking if a cemetery has a dedicated perpetual care fund.

CREMATION

If freeze-drying, rendering, or digging a hole in the backyard don't seem like good options, what to do? You might consider cremation. Cremation represents the professionalizing of the death industry for pets: the number of pet crematories is growing rapidly in response to increasing demand, and my impression is that cremation is now more common than burial in a yard or cemetery. The cremation process is becoming standardized, with various ancillary businesses providing support (pet-specific incinerators, urns in various sizes). Vets typically serve as the death middlemen. They take the bodies from owners, and then—in a process that generally remains opaque—pass them on to crematory operators. If an animal is euthanized at the vet's office, a veterinary assistant will, once the animal's family is well out of sight, place the body in black trash bag and stick it in the freezer until the scheduled crematory pick-up day. Once a week, perhaps, a truck will come and fetch a load of bodies and take them to the crematory. For Ody's in-home euthanasia, the vet took his body in the back of her car, wrapped in his purple blanket, and delivered it straight to the crematory. I'm not sure, but my guess is that his body was stored in a freezer until his appointed cremation time.

There are three types of pet cremation: private, comingled, and partitioned. In a private cremation, only one animal's body is in the oven. During a partitioned cremation, multiple animals may be in the incinerator at the same time, but they are separated so that the remains from each can be collected separately. Some "active comingling" of remains is unavoidable. Communal cremation is the burning of several animals at once, without any form of separation. Coleen Ellis told me that pet owners are often confused, and occasionally misled, about what kind of cremation their animal receives. They may ask for their animal's remains to be returned and assume that this means the animal received a private cremation, when in fact it might have been a partitioned cremation. The cremains may be mostly their animal, but active comingling means that the cremains will also include little tiny bits of other pets. Even with truly private cremations, some residual mixing—what the industry calls "unavoidable incidental comingling"—of remains will occur, since it is near impossible to remove every speck of material from the oven in between cremations.

As you know by now, cremation is how I chose to handle Ody's body. (I know "handle" is a euphemism, but I don't like the way "dispose of" sounds.) I had planned to visit various crematoria and cemeteries prior to Ody's death, to make an informed choice about what would happen to his body, were we unable to bury him at our cabin. But as it happened, I was ill-prepared; Ody's death came well before I was ready.

It was with some trepidation then that I called Pennylane Pet Cremation Services to ask if I could visit. I wanted to see where Ody's body had taken its final journey, but I was also afraid to look. The man who answered the phone was the owner himself, Chuck Myers. He was as nice as could be and told me to come out anytime and he would show me around. I set an appointment for the next day.

The crematorium is on his home property in Mead, on the flat farm country that fans out from the base of the Rockies. Exiting off Interstate 70, I drove down a rural road for about a mile, and there was the Pennylane sign and, behind it, the driveway leading to the neat red buildings of the farm. As I stepped out of my car, a black lab jumped out of his fenced yard and ran to say hello, wiggling uncontrollably and sniffing me up and down and all around. After the greeting, he turned his attention to chasing his tail, biting the tip when he could catch it. Chuck walked over and shook my hand in warm greeting. Things were off to a good start.

Unprompted, Chuck began telling me about Pennylane—what was where, how things worked, and what exactly was done with Ody's body. He explained, first, why he was wearing a suit and tie: his other job is running a human mortuary and funeral home. This, in fact, is how he got into the pet crematory business. He would get calls all the time from people who wanted to have their pets cremated. But of course, he said, they couldn't use the same oven that they use for people and he always had to turn them down. So he started the pet crematorium on the side. He saw an unmet need and decided to fill it. And business has been very good.

The crematorium itself is in their big red barn, retrofitted for the purpose. It is spacious and attractive—sparkling clean and very organized. The oven was running when I came, making a low hum. The air felt heavy and had an acrid odor, and I started to feel a bit queasy after twenty minutes in the room. Chuck told me all about how the oven works; how there is a U-shaped space under the platform where the body is and there are baffles down there breaking up the smoke. 1680 degrees Fahrenheit is the temperature that the EPA requires them to use, to minimize pollution. If the oven runs too cool, there will be more smoke and more discharge from the pipes, and the Environmental Protection Agency will levy fines. Complicated piping snakes up out of the oven and through the roof.

The oven is the exact same one they use for cremating people, only designed for smaller bodies. It can hold up to 750 pounds, so group cremations might be made up of seven to ten animals, depending on their size. He said they get all kinds of critters in addition to cats and dogs: they've had goats, rats, snakes, birds. They even had a request for cremation of an alpaca once but had to turn it down. The oven can't handle anything that big. He told me that to do large animals they would have to quarter them first, and they just didn't want to do this kind of work. An average burning takes between one to three hours (one hour for a dog like Ody; three maybe for a group cremation). He said some families will come and just sit by the oven for the whole time their animal is in there. I feel a stab of guilt when he says this. I should have been with Ody for this final part of his journey.

Next he showed me the apparatus called "the processor." I already knew that cremains are not ashes from the body but are really just crushed up bones—but I'm still a bit unsettled by the idea of Ody's bone being blenderized. Chuck said that the bones are very recognizable when they come out of the oven; you can tell that there is a skull, the long leg bones, the shorter leg bones. The bones are fed into an industrial size metal blender and crushed into a powder. He showed me a bag of remains that was about to be closed and put in an urn. Most of the powder was very fine, like ash you would clean out of your fireplace, but there were some slightly larger pieces, about the size of small pebbles, and these were recognizable as little pieces of bone. I tried to maintain a detached curiosity.

Along the back of the worktable were stacks of urns made of very light wood, like the balsawood that we used for making toy airplanes when I was a kid, which they paint with a light stain. There were big urns and teeny ones. I recognized Ody's urn, which is midsize—about 6 inches high by 4 wide. The teeny ones, Chuck said, would be for cats and maybe dachshunds. The big ones were for dogs seventy pounds and upward, such as Labradors or St. Bernard's. The urns are only for the ashes of animals cremated privately. The remains from the group cremations—the ashes that owners do not want returned—are spread on the Myers' property. "They never go in the trash," he assured me. They go out in the ample fields around the farm or in his wife's flower garden.

The remains are poured into a plastic bag and tightly closed with a zip tie so they won't spill out when you open the top of the urn or if you happen to drop it. Also in the urn with the remains is a small round piece of metal—an ID tag—that was burned along with each animal. This serves as a way to make sure that the crematory doesn't lose track of which remains belong to which animal. Chuck has a complicated system of numbers and tags and files to help him make sure that the right remains are matched with the right body. "My wife says I have too much paperwork," he joked. But it made him feel confident that no mistakes would be made.

Chuck also showed me the freezer where they store bodies until they are ready to cremate. We didn't look inside it. I was curious, but I didn't want to ask. He obviously felt that that the bodies deserve respect. He reluctantly admitted that the animals were "bagged" (quite literally, in plastic bags), "but only because we have to. We never treat a body like trash." Incidentally, the use of trash bags is one of the things that advocates of better animal aftercare want to change. Dr. Cooney said to me, "We *do* treat these bodies like trash. It looks like trash; we carry it like trash." Instead of black Hefty bags, we should use real body bags, which are becoming more readily available for pets.

I asked Chuck how people felt toward their pets' bodies, as compared to human bodies, since he has daily experience with both. His impression: people are concerned about their pets and very attentive and loving. He said he comes across a surprising number of people at his mortuary who will say things such as, "Can't you just hold on to mom's body until I'm back from vacation?" or who seem put out by the death of a family member. They just want the whole thing over and done with and out of their hair. Not so with pets. People are really concerned that the process is done right and that their animals are treated well, even after death.

Although many people are concerned about what happens to their animal's body, they sometimes might not like what they see. Coleen Ellis admits that there are unscrupulous providers, like the crematory operator who took a grieving family's dog, and their money, and gave them back some cremains—but never actually cremated their pet, instead leaving the carcass to rot outside his building. Kathy Cooney told me that veterinarians rarely want their own pets treated the same as the pets of their clients.

If you are concerned about what happens to your pet's body after you leave the vet's office or after the vet takes your animal from your home, she advises, ask a lot of questions. What will happen to the body? Will it be cremated or taken to a landfill or donated to a veterinary school? If you have chosen and paid for cremation, how exactly is the body handled? How it is wrapped, transported? Is it stored in a cooling room or freezer? How can I be assured that the cremains are really from my pet? Although it was uncomfortable for me to know the gory details, I also found it very helpful. I just wish I had asked these questions before Ody died, rather than after the fact.

REMEMBERING

When I first started work on this project, the world of pet memorialization seemed hopelessly cute, with an overabundance of rainbows and paw prints and sappy poems. You can purchase all manner of knickknacks to remind you of your animal: a marble paper weight with your pet's portrait, garden flagstones with your pet's name and a paw print, or a laser-engraved Christmas ornament. A company called Catty Shack Creations will weave you a handbag out of your pet's hair and you can even have your pet's cremains turned into a diamond or placed inside a jewelry cremation pendant. I used to find this kind of thing sentimental verging on creepy. But I have learned throughout this process that we each have our own way of expressing grief and love. I never expected to have a clay paw print of Ody sitting on my desk, but there it is, and it even has a couple of little hearts carved into it.

The website petloss.com promotes a weekly candle ceremony where everyone around the globe who has lost a pet lights a candle each Monday night in honor of the lost pet. All the animals on their Rainbow Bridge list (terminology I will explore later this chapter in the section titled "Animal Heaven") will be honored in the candle ceremony. Anyone can add their animal's name. I look through the list—which numbers easily in the thousands—and locate several other dogs named Odysseus and one named Ody. To my surprise, I find myself strangely touched by this list, and I fill out the little form with Ody's name, breed, date of birth, and date of death. Then I click "submit."

Bereavement books and websites promote memorials as a way to cope with the loss of a pet and to work through the grieving process. Suggestions include having a memorial service with friends and family, scattering a pet's remains in a meaningful place, creating some kind of scrapbook, writing a song or poem, and lighting a memorial candle. Other ideas that I like are making a memorial donation to a local shelter or rescue group or planting tulips or daffodils that will bloom year after year and remind you of your pet. One of the most touching stories of memorialization was told to me by Coleen Ellis. A client of hers came up with The Washing of the Bowls ceremony. As a way to ritualize the death of his cat Bingo, the man ended Bingo's memorial service with a symbolic final washing of Bingo's food dish.

Another nice way to honor the dead is with an obituary. Mainstream newspapers rarely include obits for animals, unless they are famous. But the magazine *Animal People* sometimes includes an animal obituaries section. In the January/February 2011 issue they included obits for Old Man, a thirty-two-year-old naked mole rat, Na'au, a California sea lion, Splash, Senator Edward Kennedy's Portuguese water dog, and Rebecca, matriarch of the Asian elephants at the Performing Animal Welfare Society's ARK sanctuary. Kathy Cooney's Home-to-Heaven Euthanasia service runs a kind of online obituary. She posts the names of animals who have died on a special webpage. Ody's name appeared about a week after he died and stayed up for seven days. It said "Ody Madden, Longmont." Cooney's service also sent a sympathy card, as did Ody's regular vet office and his mobile vet. These small touches make a difference because they remind me that many people are sorry to see Ody go.

My daughter memorializes her dead rats by placing a stone over their graves. The stones are carefully chosen because they resemble the deceased in some way—color, texture, feel, overall gestalt—I'm not really sure what it is. When I try to help pick the perfect rock and say "what about this pretty grey one?" she'll inevitably reject my choice. "No, Mom," she'll say in a scandalized voice. "That doesn't look like GooGoo at all." For Ody, she created a movie on her computer, with one picture of Ody fading into another. And she wrote the long eulogy for him that appeared earlier here, in "The Ody Journal."

As I search around in my mind for how best to memorialize Ody I have a sudden dawning of the completely obvious: I have been writing his memorial for two years. This has been my eulogy, my panegyric, my ode to a red dog. And, unbeknownst to me, I have surrounded myself with keepsakes.

Ody's oatmeal dog bed is still under the piano and his dog dish, which I cannot bear to put away, still sits on the kitchen shelf. The black collar with red and orange and green dog-bone shapes hangs off the lamp in my office, and I've left another collar hanging on a rack by the back door. His ashes sit expectantly on the corner of my desk, in their little round casket. Above my desk, I've hung a photo of Ody taken less than an hour before he died. It is a side shot of his face. His red is very deep, and the white surrounding his muzzle is muted. He looks as if he is focusing hard on some far off place.

GRIEF

People who have been touched by the loss of an animal, or who have counseled people who are grieving, realize that the grief we experience can be every bit as sharp and painful and lasting (and pathological) as the grief over the loss of a person. For some, it can be even worse. Those dismissive of animals also tend to be dismissive of this grief; they may say "it's just a dog" and "you can get another." But these are people whose lives have not been touched deeply by an animal and who don't know any better.

Of all the areas of animal death, bereavement and pet loss have perhaps been given the most careful attention. This isn't to say that everyone understands or sympathizes with pet loss, but there are plentiful resources available for those who are grieving over the death of a pet, from books and websites, to professional bereavement counselors, to online chat groups.

People are often surprised by the power of their grief for an animal. Coleen Ellis told me that people often experience more powerful feelings of grief for a lost pet than for humans they have lost. Sometimes people try to rank their grief and feel guilty if they feel more sadness at the loss of their pet than they do when their mother or husband dies. But grief for our animals is more straightforward, Ellis says, and thus in some ways more pure and condensed than our grief for people. With animals, there is no emotional baggage since their love for us is unconditional, as is our love for them. There is just pure grief. I'm not sure this is true, at least for me. Although I agree that our love for our animals may be less convoluted that our love for humans, it isn't, in my experience, without its complications. I'm not sure our love for animals is unconditional—we have conditions they must meet. And I'm not sure their love for us is unconditional either.

We are instruments in their band, just as they are instruments in ours. As much as I loved Ody, he was an albatross around my neck, a near constant source of worry. He didn't make life easy for me, and I always told him, "It's a good thing you are so lovable or you'd be in big trouble." So mixed with my grief is a certain feeling of release, of relief.

Grief is a long cycle, beginning with anticipatory grief, and moving through profound grief at the moment of death, and into the process of bereavement after the fact. For me, the anticipatory grief—the sense of impending loss—was by far the worst stage. I mourned for Ody long before he even came close to death. The moment of his death was sharp and painful—the kind of grief that makes you feel as if you're drowning. But that didn't last more than a few hours. Now, for the most part, the grief sits quietly in the back of my mind. When I see something that reminds me of Ody, when I look at his picture above my desk or find red hairs stuck to an old blanket in the garage or when I see another Vizsla I feel a stab of that sharp grief. But otherwise, it is just a soft presence.

ANIMAL HEAVEN

In Garth Stein's novel *The Art of Racing in the Rain*, the narrator tells us that in Mongolia, dogs are buried high in the hills so no one walks on the graves. The master whispers into his dog's ear his wishes that the dog will return as a man in his next life. The dog's tail is then chopped off and placed beneath his head. A piece of meat is put into his mouth, as sustenance for the long journey ahead. The dog's soul is now free, and he can run across the high Mongolian desert for as long as he wants. Running, always running . . . this is what I imagine dogs are doing in heaven.

In *Dog Heaven*, by Cynthia Rylant, we are told that, "when dogs go to heaven, they don't need wings because God knows that dogs love running best. He gives them fields. Fields and fields and fields. When a dog first arrives in heaven, he just runs." When he is done running, the dog is petted and reminded how good he is. This sounds just right to me.

One of the most enduring pieces of pet death folklore is the Rainbow Bridge. This prose poem pops up with alarming frequency in pet-loss books and websites. It is usually credited to "Anonymous," perhaps because no one can agree on who actually wrote it—and because it has been copied so many times on so many Internet sites and in so many pet bereavement pamphlets and books. The poem may have been written between 1980 and 1992 by Paul C. Dahm, William N. Britton, or Wallace Sife, but no one seems certain. Many sources suggest that the original inspiration for the Rainbow Bridge story is the ancient Norse legend of the Bifröst Bridge. Those of you familiar with Marvel Comics and Thor, the god of Thunder, will recognize this Bifröst Bridge as the shimmering bridge that connects Asgard to Earth and allows the gods to travel back and forth.

According to the poem, when companion animals die they go to a place called the Rainbow Bridge, right across the road from heaven. Next to the Bridge is a lovely meadow where our animals, restored to health and relieved of all pain and suffering, frolic and chase balls and rabbits and eat as many bones or sprigs of catnip as they want. Here in the meadow they run around and wait for us, their human companion, for however long it takes. When finally they see us coming, they bound toward us (running, once again), and after our joyful reunion we walk across the Bridge hand-in-paw, and enter heaven together.

In her book *Heaven*, Lisa Miller explores our enduring fascination with the afterlife. According to Miller, roughly 80 percent of Americans say they believe in some kind of heaven. What about pet heaven? Apparently one of the most frequently asked questions online is, "Do pets go to heaven?" It isn't surprising that the answers provided are all over the board, but it is startling to see the vehemence with which animals either are or are not permitted past the pearly gates. I am delighted to report that Billy Graham has said, in an interview, that pets do indeed go to heaven—though only because our happiness couldn't be complete without them.

Animal heaven poses some difficult theological questions. What happens if I have had a large number of pets throughout my life? Will they all run to meet me at the Rainbow Bridge? Will the cats and dogs fight? Who will walk on my right side? Theologians as venerable as Saint Augustine and Thomas Aquinas have deliberated over the heavenly estate of animals: do animals have souls, do animals go to heaven or hell or somewhere else altogether, and how many animal angels will fit on the head of a pin? ("Who knoweth the spirit of man whether it goeth upward, and the spirit of the beast whether it goeth downward to the earth?" Eccles. 3:21) Not surprisingly, there is very little consensus.

Although the Rainbow Bridge is a sweet story, it strikes me as rather self-serving. Do our animals really spend their time in the meadow just waiting for us, or might they not have their own, inscrutable business? Perhaps,

I think, they might stop by the bridge for a quick greeting and a hurried scratch behind the ears and then be off. Perhaps they have their very own animal heaven where human beings are not allowed.

BORDER CROSSINGS

In *Merle's Door*, Ted Kerasote offers this piece of dog history, from an Arctic culture: "In the very earliest time when both people and animals lived on earth a person could become an animal if he wanted to and an animal could become a human being. Sometimes they were people and sometimes animals and there was no difference." The human-animal boundaries are blurred, as perhaps they are again when we cross into that final frontier.

All is not black and white—not death and life, not animal and human. Death is a crossing over between life and death, physical and spiritual, animal and human. Dying takes us out to some wild place where the form of the animal dissolves, where animal and human bleed into one, where we can no longer hold up the edifice of difference. We all become liminal creatures. "With all eyes the creature sees the Open," writes Rilke in his *Eighth Elegy*. Perhaps this liminal space is the Open into which the animal stares and from which humans alone turn away out of fear and cowardice.

What do animals know that we do not? Does what we label their simplicity in fact amount to a wisdom that we cannot even fathom? In a poem about dogs at an Oklahoma animal shelter—dogs whose lives will soon be extinguished—Douglas Goetsch writes,

> When dogs gaze out in the same direction
> as you, sniffing the wind, they seem to know
> the future.

INTO THE OPEN

Aspen Brook meanders down the valley between Lily Mountain and Estes Cone. If you walk from our cabin up the old teahouse road, which sidles along Aspen Brook toward Lily Lake, you will arrive after about ten minutes of easy strolling at a place known to our family as the Wigwam Meadow. When I was young, an old wooden sign with "Wigwam" painted in irregular black letters was nailed to a tall Ponderosa, on the east side of the meadow. The tree has since died, the sign has disappeared, and what remains is a rotting stump. This stump seems to call with soft, craggy voice, "You have arrived. You are here."

It is a small meadow, perhaps only a hundred paces from one end to the other. On one side, the meadow is bordered by the rutted and uneven road. Ponderosa pines, blue spruce, and Douglas firs form an arc around the other three sides. One large rock, splotched with lichens—the subdued green *Xanthoparmelia*, the salmon-orange *Rhizoplaca*—forms the meadow's highest point. The meadow is carpeted with kinnikinnick and all manner of mountain grasses, which shift colors with the movement of the sun and make a soft whishing sound in the wind. In spring, Pasque flowers burst unexpectedly from the late snow on the side of the road, and higher up in the meadow the diminutive spring beauties peek out at the sun. In summer, the meadow is flecked with dainty white chickweed, lavender tansy asters, and bright gold mustard flowers. The smell of prairie sage lightly flavors the air. On the other side of the road and down a short slope, the brook flows past. From this distance, its song is low and gentle as a lullaby.

My brother and I camped out in this meadow when we were little, drinking hot cocoa from a thermos and watching meteors fall like dying embers through the night sky. As an adult, I have walked here thousands upon thousands of times with Ody, watching him run crazy through the grasses after the scent of a chipmunk and, more recently, struggling up the small slope, back legs dragging, but nose held high, still alert to the meadow's olfactory delights.

This meadow is where I will spread Ody's ashes, out among the sage and kinnikinnick. For now he sits patiently on my desk, in his wooden urn decorated with a pale yellow ribbon. But when I gather up the nerve to say my last goodbye, Ody and I will take one final journey, up to the meadow. Maybe we'll go soon, while the ground is still hard with frost and forest sounds are muted by cold, or maybe we'll take this final odyssey in the spring, when life begins to stir again and the Pasque flowers are blooming.

This is my version of dog heaven: I stand in the Wigwam Meadow and I blow my horn, and Ody comes running, sleek and red, into my arms. From here, we step together into the Open.

The Ody Journal

In the first stanza of Robinson Jeffers's poem "The House-Dog's Grave," the ghost dog says to his human companion:

> I've changed my ways a little: I cannot now
> Run with you in the evenings along the shore,
> Except in a kind of dream and you,
> If you dream a moment,
> You see me there.

Ody has changed his ways a little, too. But he is still here and will be part of the geography of my life forever. He has etched himself into our house, through the stitched-up scars in the couches, the extra tall fences, the scratches in the doorframes, and the fact that every blanket and bed-spread we own has holes from Ody trying to dig his way out of anxiety's grip. And he has etched himself into my heart, as fully and painfully as any creature ever has. He has become a part of me.

Even now, I wake to phantom barks in the middle of the night, worried that Ody needs to go out or that he is out and needs to come in or that he is trapped in some strange corner of the house, back legs splayed out in an agony of helplessness. I lie still, feeling adrenaline finger its way into my veins, and I breathe shallow in the darkness, waiting for the next hoarse "arouf." But there is only stillness.

During the day, I sometimes feel Ody looking at me through the office door; I feel his eyes on me. And out of habit I turn to offer my ritual greeting. In a kind of dream, I push back my chair, step into the doorway where he waits, and take his soft red head into my hands.

Acknowledgments

I would like to thank the many people who offered their time and expertise, especially Marc Bekoff, Gail Bishop, Kathy Cooney, Robin Downing, Eric Greene, Leslie Irvine, Bernie Rollin, Alice Villalobos, and the folks at Pennylane. My heartfelt appreciation goes out to everyone at University of Chicago Press. Thanks especially to my wonderful editor, Christie Henry, to my manuscript editor and fellow dog-lover Yvonne Zipter, and to Levi Stahl and his marketing team. Special thanks go to my family and friends, who supported this project in one way and another: to my husband Chris, for his friendship and for always encouraging me to forge my own path; to my daughter Sage, for demanding that I be bold; to my parents Roger and Alexandra, for listening to all my animal stories and nurturing my life passions; to my brother Benjamin for always being a solid presence, no matter where we are. Thanks to Maya, for letting me know when the mail has arrived and for teaching me about being good-natured, and to Topaz, for keeping my feet warm under the desk and for protecting me always. And above all, thanks to Ody, for everything: for serving as my muse, for reminding me that animals live in a world as fabulous and mysterious as our own, for always overlooking my shortcomings, and for teaching me about aging and about graceful passage beyond. And, of course, for being inscrutable.

Notes

CHAPTER 1

PAGE 6   *In* Dog Years, *Mark Doty writes, "No dog has ever."* Mark Doty, *Dog Years: A Memoir* (New York: HarperCollins, 2007), 1.

PAGE 6   *Doty goes on to say, "Love for a wordless creature."* Doty, *Dog Years*, 3.

PAGE 7   *During one year, US consumers will have purchased.* American Pet Products Manufacturers' Association, "Pet Industry Statistics and Trends," http://www.americanpetproducts.org/press_industrytrends.asp.

PAGE 7   *The Humane Society of the United States estimates that six to eight million cats.* Humane Society of the United States, "Common Questions about Animal Shelters," October 26, 2009, http://www.humanesociety.org/animal_community/resources/qa/common_questions_on_shelters.html#How_many_animals_enter_animal_shelters_e.

PAGE 8   *Killing is by far the most common.* Animal Studies Group, *Killing Animals* (Champaign: University of Illinois Press, 2006). The quote is drawn from the book's description on the University of Illinois Press website: http://www.press.uillinois.edu/books/catalog/56xce4yy9780252030505.html.

PAGE 8   *Jonathan Safran Foer, for example, asserts that 96 percent.* Jonathan Safran Foer, *Eating Animals* (New York: Little, Brown, 2009), 73.

PAGE 12   *Recognizing that the ability to turn around.* Farm Animal Welfare Council, "Five Freedoms," last modified April 16, 2009, http://www.fawc.org.uk/freedoms.htm.

CHAPTER 2

PAGE 29   *In a* New York Times *story about death awareness in animals.* Natalie Angier, "About Death, Just Like Us or Pretty Much Unaware?" *New York Times*, September 1, 2008: http://www.nytimes.com/2008/09/02/science/02angi.html.

PAGE 29   *A 2009* Daily Mail *headline read, "Is This Haunting Picture."* Michael Hanlon, "Is This Haunting Picture Proof That Chimps Really DO Grieve?" *Daily Mail,* October 27, 2009: http://www.dailymail.co.uk/sciencetech/article-1223227/ Is-haunting-picture-proof-chimps-really-DO-grieve.html.

PAGE 29   *Gana, an eleven-year-old gorilla at the Munster Zoo in Germany.* Marcus Dunk, "A Mother's Grief: Heartbroken Gorilla Cradles Her Dead Baby," *Daily Mail,* August 19, 2008: http://www.dailymail.co.uk/sciencetech/article-1046549/A-mothers-grief-Heartbroken-gorilla-cradles-dead-baby.html.

PAGE 30   *Goodall says, "Never shall I forget watching."* Jane Goodall, *Through a Window: My Thirty Years with the Chimpanzees of Gombe* (Boston: Houghton Mifflin, 1990), 196–97.

PAGE 30   *"A graylag goose that has lost its partner."* Konrad Lorenz, *Here Am I—Where Are You? The Behavior of the Greylag Goose* (New York: Harcourt), 251.

PAGE 30   *Zoologist Iian Hamilton believes that elephants.* See, e.g., Ian Douglas-Hamilton et al. "Behavioural Reactions of Elephants towards a Dying and Deceased Matriarch," *Applied Animal Behaviour Science* 100 (2006): 87–102.

PAGES   *"Even bare, bleached old elephant bones will stop."* Cynthia Moss, *Elephant*
30-31   *Memories* (Chicago: University of Chicago Press, 2000), 270–71.

PAGE 31   *A study of tool use in African elephants.* Suzanne Chevalier-Skolnikoff and Jo Liska, "Tool Use by Wild and Captive Elephants," *Animal Behaviour* 46, no. 2 (1993): 209–19.

PAGE 31   *And biologist Joyce Poole writes of elephants, "I have observed."* Joyce Poole, *Elephants* (Stillwater, MN: Voyageur Press, 1997), 12.

PAGE 31   *According to a report by the Cornell Lab of Ornithology.* Janis L. Dickinson and Miyoko Chu. "Animal Funerals," *BirdScope* 21, no.1 (2007): http://www.birds .cornell.edu/Publications/Birdscope/Winter2007/animal_funerals.html.

PAGE 31   *Ethologist Marc Bekoff observed the following behavior.* Reported in "Magpies 'Feel Grief and Hold Funerals,'" *Daily Telegraph,* October 21, 2009: http://www. telegraph.co.uk/earth/wildlife/6392594/Magpies-feel-grief-and-hold-funerals .html.

PAGE 31   *"There can be no doubt," proclaims Michael Fox.* Michael W. Fox, *Dog Body, Dog Mind: Exploring Canine Consciousness and Total Well-Being* (Guilford, CT: Lyons Press, 2007), 86.

PAGE 32   *"Animals, in their grief, and in their longing."* Fox, *Dog Body, Dog Mind,* 81.

PAGE 32   *The Companion Animal Mourning Project.* International Association for Animal Hospice and Palliative Care, "Canine Grief: 'Do Dogs Mourn?'" http://

iaahpc.org/index.php/for-pet-parents/helpful-articles/item/8-canine-grief-%E2%80%93-do-dogs-mourn.

PAGE 34   *We are talking to Mr. Beau.* Doty, *Dog Years*, 147.

PAGE 35   *"Miller et al., Langer, and others state."* Donald R. Griffin, *The Question of Animal Awareness* (New York: Rockefeller University Press, 1981), 104–5.

PAGE 36   *Oscar's story gained unusual credibility when Dr. David Dosa.* David M. Dosa, "A Day in the Life of Oscar the Cat," *New England Journal of Medicine* 357, no. 4 (2007): 328–29.

PAGE 37   The Greyfriars Bobby website can be found at http://www.greyfriarsbobby .co.uk/.

PAGE 38   *Veterinarian Klaus Mueller said of Barnaby.* www.thefreelibrary.com/The+ bull+in+mourning%3B+Farmer's+'pet'+holds+a +two-day+vigil+at+his . . . -a0114754353. This link is no longer active.

CHAPTER 3

PAGE 57   *"Aging is not so much a direct cause."* André Klarsfeld and Frédéric Revah, *The Biology of Death: Origins of Mortality*, trans. Lydia Brady (Ithaca, NY: Cornell University Press, 2004), 36.

PAGE 59   *"Until recently, people believed that wild animals."* Anne Innis Dagg, *The Social Behavior of Older Animals* (Baltimore: Johns Hopkins University Press, 2009), 1–2.

PAGES   *For example, a recent study by Karen McComb.* Karen McComb et al, "Leadership
59-60   in elephants: the adaptive value of age," *Proceedings of the Royal Society* B 278 (2011): 3270–76.

PAGE 60   *Within the population of companion animals, the elderly is the fastest.* Chris C. Pinney, *The Complete Home Veterinary Guide*, 3d ed. (New York: McGraw-Hill, 2004), 641.

PAGE 60   *There are about seventy-eight million companion dogs.* American Pet Products Manufacturers' Association, "Pet Industry Statistics and Trends."

PAGE 62   *The brains of dogs with senile dementia.* David Taylor, *Old Dog, New Tricks: Understanding and Retraining Older and Rescued Dogs* (Buffalo, NY: Firefly Books, 2006), 76.

PAGE 62   *"Although they are experienced in life and set in their ways."* Taylor, *Old Dog, New Tricks*, 76.

PAGE 69   *A* New York Times *article about joint replacements.* Vincent M. Mallozzi, "Joint Replacements Keep Dogs in the Competition," *New York Times,* January 17, 2010, D1.

PAGE 70   *Indeed, a recent survey reported that 81 percent.* Stanley Coren, "Do We Treat Dogs the Same Way as Children in Our Modern Families?" *Canine Corner: The Human-Animal Bond* (blog), *Psychology Today,* May 2, 2011: http://www .psychologytoday.com/blog/canine-corner/201105/do-we-treat-dogs-the-same-way-children-in-our-modern-families.

PAGE 75   *According to Fred Metzger, a veterinarian who specializes in senior animals.* Grey Muzzle Organization, "Old Dogs and Animal Shelters," http://www .greymuzzle.org/PDF/OldDogsandAnimalShelters.aspx.

PAGE 76   *Although the relationship between pet ownership and well-being.* For a review of the literature, see Miho Nagasawa and Mitsuaki Ohta, "The Influence of Dog Ownership in Childhood on the Sociality of Elderly Japanese Men," *Animal Science Journal* 81 (2010): 377–83.

PAGE 77   *The* New York Times *recently ran a piece about the Thoroughbred.* Joe Drape, "Ex-Racehorses Starve as Charity Fails in Mission to Care for Them," *New York Times,* March 17, 2011: http://www.nytimes.com/2011/03/18/sports/18horses.html.

CHAPTER 4

PAGE 90   *"Animals subjected to potentially painful or stressful."* Cornell University Institutional Animal Care and Use Committee, "Guidelines for Assigning Animals into USDA Pain and Distress Categories," December 2009, http://www.iacuc .cornell.edu/documents/IACUC009.01.pdf.

PAGES   *According to the National Research Council's Committee.* National Research
91–92   Council, Committee on Pain and Distress in Laboratory Animals, *Recognition and Alleviation of Pain and Distress in Laboratory Animals* (Washington, DC: National Academy Press, 1992), 5 (http://www.ncbi.nlm.nih.gov/books/ NBK32656/).

PAGE 92   *As the International Association for the Study of Pain says.* International Association for the Study of Pain, "IASP Taxanomy," last updated July 14, 2011, http://www.iasp-pain.org/AM/Template.cfm?Section=Pain_Defi . . . isplay .cfm&ContentID=1728.

PAGE 92   *The National Research Council tells us that "the encoding."* National Research Council, *Recognition and Alleviation,* 5.

PAGE 96   *It is nice that the government recognizes that animal suffering.* National Research Council, *Recognition and Alleviation,* 85.

PAGE 96   *Animal behaviorist Marian Stamp Dawkins defines suffering as.* Marian Stamp Dawkins, "The Scientific Basis for Assessing Suffering in Animals," in *In Defense of Animals*, ed. Peter Singer (New York: Basil Blackwell, 1985), 49.

PAGE 97   *"To bridge that gap, we each have to make."* Marian Stamp Dawkins, "The Science of Suffering," in *Mental Health and Well-Being*, ed. McMillan, 48.

PAGE 97   *Jaak Panksepp, a neurobiologist who studies the physiological roots.* Jaak Panksepp, "Affective-Social Neuroscience Approaches to Understanding Core Emotional Feelings in Animals," in *Mental Health and Well-Being*, ed. McMillan, 61.

PAGES 99–100   *Our scientist notes that the adrenal response to stress.* Jonathan Balcombe, *Second Nature: The Inner Lives of Animals* (New York: Palgrave Macmillan, 2010), 17.

PAGE 100   *Our veterinarian suggests that animals may suffer more severely.* Bernard E. Rollin, *Science and Ethics* (New York: Cambridge University Press, 2006), 236. Referring to Kitchell and Guinan (1989).

PAGE 100   *And finally, our philosopher points out, "If animals are indeed..."* Rollin, *Science and Ethics*, 238.

PAGE 102   *He quotes a dean of a vet school saying, "Anesthesia and analgesia."* Bernard E. Rollin, *The Unheeded Cry: Animal Consciousness, Animal Pain, and Science*, expanded ed. (Ames: Iowa State University Press, 1989), 117.

PAGE 103   *A decade ago, Dr. Sheilah A. Robertson, opening speaker.* R. Scott, Nolen, "Silent Suffering: AVMA Animal Welfare Forum addresses pain management in animals," *American Veterinary Medical Association News*, December 15, 2001: http://www.avma.org/onlnews/javma/dec01/s121501c.asp.

PAGE 104   *Human Rights Watch published a report in 2009.* Human Rights Watch, "Please, Do Not Make Us Suffer Any More... : Access to Pain Treatment as a Human Right," 2009, http://www.painandhealth.org/sft241/hrw_please_do_not_make_us_suffer_any_more.doc1.pdf.

PAGE 105   *He goes on: "Veterinary use of analgesics appeared."* Kevin Stafford, *The Welfare of Dogs* (Dordrecht: Springer, 2006), 124.

PAGE 106   *For example, based on "pain scores derived from behavior-based."* Marijke E. Peeters and Jolle Kirpensteijn, "Comparison of Surgical Variables and Short-Term Postoperative Complications in Healthy Dogs Undergoing Ovariohysterectomy or Ovariectomy," *Journal of the American Veterinary Medical Association* 238 (2011): 189–94.

PAGE 107   *Of those who are treated, he estimates that many will be treated ineffectively.* Stafford, *Welfare of Dogs*, 126.

PAGE 107   *The National Research Council concludes that "there are no."* National Research Council, *Recognition and Alleviation*, 33.

PAGE 107   *Robin Downing believes that one of the most important challenges.* Robin Downing, "Pain Management for Veterinary Palliative Care and Hospice Patients," in *Palliative Medicine and Hospice Care*, ed. Tami Shearer, Veterinary Clinics of North America: Small Animal Practice, vol. 41, no.3 (Philadelphia: W. B. Saunders, 2011), 533.

PAGE 108   *The International Veterinary Academy for Pain Management lists the following.* International Veterinary Academy of Pain Management, "Treating Pain in Companion Animals," 2, http://www.carolstreamah.com/ce/ivapm_ petownerinfsheet112005.pdf

PAGE 108   *The factsheet "How to Recognize Pain in Your Dog."* National Academy of Sciences, "How to Recognize Pain in Your Dog," 2010. http://dels.nas.edu/ resources/static-assets/materials-based-on-reports/special-products/dog_ factsheet_final.pdf.

PAGE 111   *Robin Downing uses the image of a pain-management pyramid.* Downing, "Pain Management," 536ff.

PAGE 112   *Animals most certainly experience pleasure, as biologist Jonathan Balcombe.* Jonathan Balcombe, *The Exultant Ark: A Pictorial Tour of Animal Pleasure* (Berkeley: University of California Press, 2011), and *Pleasurable Kingdom: Animals and the Nature of Feeling Good* (New York: Palgrave McMillan, 2007).

PAGE 112   *The reason for this, veterinarian Frank McMillan speculates.* Franklin D. McMillan, "Do Animals Experience True Happiness?" in *Mental Health*, ed. McMillan, 223.

PAGE 115   *"For example," the researchers say, "if an animal is in an environment."* Michael Mendl, Oliver H. P. Burman, and Elizabeth S. Paul, "An Integrative and Functional Framework for the Study of Animal Emotion and Mood," *Proceedings of the Royal Society* B 277 (2010): 2899.

PAGE 116   *For example, a team from Newcastle found that pigs.* "Can You Ask a Pig If His Glass Is Half Full?" *Science Daily*, July 7, 2010: http://www.sciencedaily.com/ releases/2010/07/100727201515.htm.

PAGE 116   *A study by Michael Mendl (author of our mood-state essay) on optimism and pessimism.* Michael Mendl et al., "Dogs Showing Separation-Related Behavior Exhibit a 'Pessimistic' Cognitive Bias," *Current Biology* 20(2010): R839–R840.

PAGE 117   *As Mendl, Burman, and Paul note, "There is a whole industry."* Mendl, Burman, and Paul, "Integrative and Functional Framework," 2895.

PAGE 119   *Over the past decade, the field of animal personality research.* See, e.g., Sam

Gosling's Animal Personality Lab website, "Animal Personality," last modified October 18, 2011, http://homepage.psy.utexas.edu/homepage/faculty/gosling/animal_personality.

CHAPTER 5

PAGE 131 *As Kathryn Marrochino says, quoting Tom Wilson.* Kathryn Marrochino, "In the Shadow of a Rainbow: The History of Animal Hospice," in *Palliative Medicine*, ed. Shearer, 492.

PAGE 135 *The association has taken an interest.* American Association of Human-Animal Bond Veterinarians, "End of Life Hospice Care," 2010, http://aahabv.org/index.php?option=com_content&view=article&id=71&Itemid=93.

PAGE 135 *One veterinarian I interviewed estimated.* Kathy Cooney, personal communication, September 2, 2011.

PAGE 135 *Another estimated that the number of animals treated annually.* Amir Shanan, e-mail message to author, September 26, 2011.

PAGE 137 *Euthanasia rates at Angel's Gate are reportedly very low.* Marocchino, "In the Shadow of a Rainbow," 483.

PAGE 141 *"In virtually all cases of progressive incurable illness."* Johnny Hoskins, *Geriatrics and Gerontology of the Dog and Cat* (Philadelphia: W. B. Saunders, 2003), 10.

PAGE 143 *McMillan notes that within human medicine.* Franklin D. McMillan, "The Concept of Quality of Life in Animals," in *Mental Health*, ed. McMillan, 193.

PAGE 149 *A hospice flyer from Colorado State University's Argus Institute.* Leah Berrett, Carol Borchert, and Laurel Lagoni, *What Now? Support for You and Your Companion Animal*, 2d ed. (Fort Collins: Argus Institute, College of Veterinary Medicine and Biomedical Sciences, Colorado State University, 2009), 5.

PAGE 150 *According to a survey by the American Animal Hospital Association.* Canadian Veterinary Medical Association, "Pet Owners Let Love Rule," *Canadian Veterinary Journal* 43 (2002): 344: http://www.ncbi.nlm.nih.gov/pmc/articles/PMC339263/.

PAGE 152 *According to the Humane Society of the United States, dog owners spend.* Humane Society of the United States, "U.S. Pet Ownership Statistics," August 12, 2011, http://www.humanesociety.org/issues/pet_overpopulation/facts/pet_ownership_statistics.html.

PAGE 152 *To put things in perspective, according to the Kaiser Family Foundation.* Kaiser Family Foundation, "Health Care Spending in the United States and Selected

OECD Countries," April 28, 2011, http://www.kff.org/insurance/snapshot/chcmo103070th.cfm.

PAGE 152 *And, for yet more perspective, the average US consumer spends $457.* Visual Economics, "How the Average U.S. Consumer Spends Their Paycheck," 2009, http://www.visualeconomics.com/how-the-average-us-consumer-spends-their-paycheck.

CHAPTER 6

PAGES
167–68
Jerrold Tannenbaum, a professor of veterinary ethics. Jerrold Tannenbaum, *Veterinary Ethics: Animal Welfare, Client Relations, Competition and Collegiality*, 2d ed. (St. Louis: Mosby, 1995), 342.

PAGE 173 *Euthanizing agents cause death by three basic mechanisms.* American Veterinary Medical Association, *AVMA Guidelines on Euthanasia* (2007), 5, http://www.avma.org/issues/animal_welfare/euthanasia.pdf.

PAGES
174–75
Fatal-Plus   Indications: For fast and humane euthanasia. Drug Information Online, "Fatal-Plus Solution," http://www.drugs.com/vet/fatal-plus-solution.html.

PAGE 177 *Apparently not, and as a case in point consider the story of Mia.* Patty Khuly, "Sometimes They Come Back: A Not-So-Euthanized Dog's Tale Goes Viral, October 14, 2010, http://www.petmd.com/blogs/fullyvetted/2010/oct/dead_dog_walking.

PAGE 178 *Cooney warns practitioners that the pet.* Kathleen Cooney. *In-Home Pet Euthanasia Techniques* (Loveland, CO: Home to Heaven, 2011), 93.

PAGE 178 *Cardiac electrical activity, she continues.* Cooney, *In-Home Pet Euthanasia*, 94.

PAGE 178 *Cooney outlines some of the most common side-effects.* Cooney, *In-Home Pet Euthanasia*, 95.

PAGE 179 *Bernie Rollin offers a few examples in his* Introduction to Veterinary Medical Ethics. Bernard Rollin, *Introduction to Veterinary Medical Ethics* (Hoboken, NJ: Wiley-Blackwell, 2006), 115, 127, 135.

PAGE 180 *Kathy Cooney writes, "If you provide this service long enough."* Cooney, *In-Home Pet Euthanasia*, 7.

PAGE 181 *Behavioral issues are thought to be the leading cause.* Stephen R. Lindsay, *Handbook of Applied Dog Behavior*, vol. 1, *Adaptation and Learning* (Ames: Iowa State University Press, 2000), 370.

PAGE 182 *I cannot verify how often people attempt euthanasia at home.* Sharon L. Peters,

"Do-It-Yourself Animal Euthanasia Is NOT Recommended," *USA Today*, March 26, 2009: http://www.usatoday.com/news/nation/2009–05–26-euthanasia-side_N.htm.

PAGES 184–85 *According to the most recent data, about 1.5 million dogs.* "2011 Shelter Data Update," *Animal People Magazine*, July–August 2011, 14: https://acrobat.com/SignIn.html?d=3HWraRZtdaBBjWKSsSU1ha.

PAGE 185 *The institute notes that "problems stemming."* International Institute for Animal Law, "Humane Euthanasia of Animals," http://www.animallaw.com/Humaneeuthanasia.htm.

PAGE 185 *Even with EBI, says Fakkema.* Doug Fakkema, telephone interview by author, August 18, 2011.

PAGE 186 *To challenge this, Fakkema compiled a detailed cost analysis matrix.* Doug Fakkema, "EBI Cost Analysis Matrix 2009," and "Carbon Monoxide Cost Analysis Matrix 2009," http://www.americanhumane.org/assets/pdfs/animals/adv-co-ebi-cost-analysis09.pdf.

PAGE 187 *It is worth noting some confusion and inconsistency in use of the terms.* See Jones and Gosling (2005) on the use of these terms in relation to temperament testing.

PAGE 187 *Her website tells us, "Sue's nationally known temperament test."* "Assess-a-Pet™," 2006, http://www.suesternberg.com/03programs/03assessapet.html.

PAGE 188 *Although many shelters swear by their methods, the validity.* Amanda C. Jones and Samuel D. Gosling, "Temperament and Personality in Dogs (*Canis familiaris*): A Review and Evaluation of Past Research," *Applied Animal Behavior Science* 95(2005): 1–53.

PAGES 188–89 *The shelter worker in charge of euthanizing animals.* Marc Lacey, "Afghan Hero Dog Is Euthanized by Mistake in U.S.," *New York Times*, November 18, 2010: http://www.nytimes.com/2010/11/19/us/19dog.html.

PAGE 189 *People for the Ethical Treatment of Animals is also pro-euthanasia.* People for the Ethical Treatment of Animals, "Euthanasia: The Compassionate Option," http://www.peta.org/issues/Companion-Animals/euthanasia-the-compassionate-option.aspx.

PAGE 189 *According to Winograd, 90 percent of all.* Nathan J. Winograd, *Redemption: The Myth of Pet Overpopulation and the No Kill Revolution in America* (Santa Clara, CA: Almaden Press, 2009), x–xi.

PAGE 190 *"People need to see what happens here."* "Miami-Dade Doggy Death Row Grows Overcrowded," 2010, http://cbs4.com/pets/Miami.Dade.Pet.2.1767060.html; this link no longer works.

PAGE 191   *The AVMA Guidelines on Euthanasia even warn.* American Veterinary Medical Association, *AVMA Guidelines on Euthanasia* (2007), 4, http://www.avma.org/ issues/animal_welfare/euthanasia.pdf.

PAGE 191   *A study out of North Carolina found that shelter workers.* Charlie L. Reeve et al., "The Caring-Killing Paradox: Euthanasia-Related Strain among Animal-Shelter Workers," *Journal of Applied Social Psychology* 35 (2005): 119–43.

PAGE 191   *Drug overdose is the most common method of suicide.* Tom Watkins, "Paper Delves into British Veterinarians' High Suicide Risk," *CNN World.* May 26, 2010, http://current.com/http://www.cnn.com/2010/WORLD/europe/03/26/ england.veterinarians.suicide/index.html?hpt=T2.

PAGE 191   *On the human costs of euthanasia, Krista Schultz writes.* Krista Schultz, "An Emerging Occupational Threat? Study Seeks Reasons for High Suicide Rate among Veterinarians," *DVM Newsmagazine,* May 1, 2008: http:// veterinarynews.dvm360.com/dvm/Veterinary+business/An-emerging-occupational-threat/ArticleStandard/Article/detail/514794.

PAGE 192   *Tom Watkins, reporting for CNN on the British study on veterinary suicide.* Watkins, "Paper Delves." The original paper is by D. J. Bartram and D. S. Baldwin, "Veterinary Surgeons and Suicide: Influences, Opportunities, and Research Directions," *Veterinary Record* 162 (2008): 36–40.

PAGE 192   *A recent survey of US physicians' found that 69 percent.* Farr A. Curlin et al., "To Die, to Sleep: US Physicians' Religious and Other Objections to Physician-Assisted Suicide, Terminal Sedation, and Withdrawal of Life Support," *American Journal of Hospice and Palliative Care* 25 (2008): 112–20.

PAGE 192   *"This apparent lack of interest is startling," he writes.* Tannenbaum, *Veterinary Ethics,* 355.

PAGE 193   *Tannenbaum's reflections on veterinary euthanasia.* Tannenbaum, *Veterinary Ethics,* 355–56.

CHAPTER 7

PAGE 204   *For example, it can be as uncomplicated.* Tannenbaum, *Veterinary Ethics,* 344.

PAGES 206-7   *As the Colorado State University Cooperative Extension explains.* Mark Cronquist, "Livestock Mortality Management," Small Acreage Series, Colorado State University Cooperative Extension, 2007, http://www.ext .colostate.edu/sam/livestock-mortality.pdf.

PAGE 207   *The Perpetual Pet website boasts.* Perpetual Pet, "Pet Preservation through Freeze Dry Technology," http://www.perpetualpet.net/.

PAGES *The Cryonics Institute explains.* Cryonics Institute, "Cryonics: A Basic
208-9 Introduction," 2010, http://www.cryonics.org/prod.html.

PAGE 211 *Various animals throughout human history have been interred.* R. J. Losey et al.
"Canids as Persons: Early Neolithic Dog and Wolf Burials, Cis-Baikal, Siberia,"
Journal of Anthropological Archaeology 30, no. 2 (2011), 174–89.

PAGE 211 *Some human graves in the area contain "canid remains . . ."* Losey et al., "Canids as
Persons," 174.

PAGE 221 *According to Miller, roughly 80 percent of Americans.* Lisa Miller, "We Believe in
Heaven, but What Is It?" *Washington Post*, June 7, 2007: http://newsweek
.washingtonpost.com/onfaith/guestvoices/2007/06/we_believe_in_heaven_
but_what.html.

PAGE 222 *In Merle's Door, Ted Kerasote offers this piece of dog history.* Ted Kerasote, *Merle's
Door: Lessons from a Freethinking Dog* (New York: Harcourt, 2007), 213. He is
quoting Marion Schwartz, *A History of Dogs in the Early Americas* (New Haven,
CT: Yale University Press, 1997), vi.

PAGE 222 *In a poem about dogs at an Oklahoma animal shelter.* Douglas Goetsch, "Different
Dogs," *New Yorker*, January 17, 2011, 60–61.

PAGE 224 *In the first stanza Robinson Jeffers's poem "The House-Dog's Grave."* Robinson
Jeffers, "The House-Dog's Grave," in *The Selected Poetry of Robinson Jeffers*, ed.
Tim Hunt (Stanford, CA: Stanford University Press, 2002), 559.

Bibliography

Agamben, Giorgio. *The Open: Man and Animal.* Translated by Kevin Attell. Stanford, CA: Stanford University Press, 2004.

Allen, Colin. "Animal Pain." *Noûs* 38, no. 4 (2004): 617–43.

Allen, Colin, Perry N. Fuchs, Adam Shriver, and Hilary D. Wilson. "Deciphering Animal Pain." In *Pain: New Essays on Its Nature and the Methodology of Its Study*, edited by Murat Aydede, 351–66. Cambridge, MA: MIT Press, 2006.

Alper, Ty. "Anesthetizing the Public Conscience: Lethal Injection and Animal Euthanasia." *Fordham Urban Law Journal*, vol. 35 (May 2008). http://www.law.berkeley.edu/clinics/dpclinic/LethalInjection/LI/documents/articles/journal/alper.pdf.

American Association of Feline Practitioners. "Veterinary Hospice Care for Cats." 2010. http://www.catvets.com/uploads/PDF/2010VeterinaryHospiceCareforCats.pdf.

American Association of Human-Animal Bond Veterinarians. "End of Life Hospice Care." 2010. http://aahabv.org/index.php?option=com_content&view=article&id=71&Itemid=93.

American Humane Association. "Animal Shelter Euthanasia." 2010. http://www.americanhumane.org/about-us/newsroom/fact-sheets/animal-shelter-euthanasia.html.

American Veterinary Medical Association. *AVMA Guidelines on Euthanasia.* 2007. http://www.avma.org/issues/animal_welfare/euthanasia.pdf.

———. "Contradictions Characterize Pain Management In Companion Animals." *Journal of the American Veterinary Medical Association*, December 15, 2001. http://www.avma.org/onlnews/javma/dec01/s121501g.asp.

———. *Guidelines for Veterinary Hospice Care.* 2011. http://www.avma.org/issues/policy/hospice_care.asp.

———. "Principles of Veterinary Medical Ethics of the AVMA." 2008. http://www.avma.org/issues/policy/ethics.asp.

———. "Silent Suffering: AVMA Animal Welfare Forum Addresses Pain Management in Animals," *Journal of the American Veterinary Medical Association*, December 15, 2001. http://www.avma.org/onlnews/javma/dec01/s121501c.asp.

———. "Veterinarian's Oath." 2010. http://www.avma.org/about_avma/whoweare/oath.asp.

Anderson, Allen, and Linda Anderson. *Saying Goodbye to Your Angel Animals*. Novato, CA: New World Library, 2005.

Anderson, James R., Alasdair Gillies, and Louise C. Lock. "*Pan* thanatology." *Current Biology* 20, no. 8 (2010): R349–R351.

Angier, Natalie. "About Death, Just Like Us or Pretty Much Unaware?" *New York Times*, September 1, 2008. http://www.nytimes.com/2008/09/02/science/02angi.html.

———. "Save a Whale, Save a Soul, Goes the Cry." *New York Times*, June 27, 2010. http://www.nytimes.com/2010/06/27/weekinreview/27angier .html?_r=1&ref=intelligence.

Animal Studies Group. *Killing Animals*. Champaign: University of Illinois Press, 2006.

Balcombe, Jonathan. "Animal Pleasure and Its Moral Significance." *Applied Animal Behaviour Science* 118 (2009): 208–16.

———. *The Exultant Ark: A Pictorial Tour of Animal Pleasure*. Berkeley: University of California Press, 2011.

———. *Pleasurable Kingdom: Animals and the Nature of Feeling Good*. New York: Macmillan, 2007.

———. *Second Nature: The Inner Lives of Animals*. New York: Palgrave Macmillan, 2010.

Bartram, D. J., and D. S. Baldwin. "Veterinary Surgeons and Suicide: Influences, Opportunities, and Research Directions." *Veterinary Record* 162 (2008): 36–40.

Bataille, Georges. "Animality." In *Animal Philosophy*, by Peter Atterton and Matthew Calarco, 33–36. New York: Continuum, 2004.

Bates, George. "Humane Issues Surrounding Decapitation Reconsidered." *Journal of the American Veterinary Association* 237, no. 9 (2010): 1024–26.

Beatson, Ruth M., and Michael J. Halloran. "Humans Rule! The Effects of Creatureliness Reminders, Mortality Salience and Self-Esteem on Attitudes toward Animals." *British Journal of Social Psychology* 46 (2007): 619–32.

Becker, Ernst. *The Denial of Death.* 1973. Reprint, New York: Free Press, 1997.

———. *Escape from Evil.* 1975. Reprint, New York: Free Press, 1985.

Becker, Geoffrey S. "Animal Rendering: Economics and Policy." Congressional Research Service Report for Congress. March 17, 2004. http://www.nationalaglawcenter.org/assets/crs/RS21771.pdf.

Becker, Marty. *The Healing Power of Pets.* New York: Hyperion, 2002.

Behan, Kevin. *Your Dog Is Your Mirror: The Emotional Capacity of Our Dogs and Ourselves.* Novato, CA: New World Library, 2011.

Bekoff, Marc. *The Emotional Lives of Animals.* Novato, CA: New World Library, 2007.

Bekoff, Marc, and Jessica Pierce. *Wild Justice: The Moral Lives of Animals.* Chicago: University of Chicago Press, 2009.

Berger, Peter L. *The Sacred Canopy.* New York: Doubleday, 1967.

Berrett, Leah, Carol Borchert, and Laurel Lagoni. *What Now? Support for You and Your Companion Animal,* 2d ed. Fort Collins, CO: Argus Institute, College of Veterinary Medicine and Biomedical Sciences, Colorado State University, 2009.

Bilger, Burkhard. "The Last Meow." *New Yorker,* September 8, 2003.

Biro, Dora, Tatyana Humle, Kathelijne Koops, Claudia Sousa, Misato Hayashi, and Tetsuro Matsuzawa. "Chimpanzee Mothers at Bossou, Guinea, Carry the Mummified Remains of Their Dead Infants." *Current Biology* 20, no. 8 (2010): R351–R352.

Bonanno, George A. *The Other Side of Sadness.* New York: Basic Books, 2009.

Boone, J. Allen. *Kinship with All Life.* New York: Harper & Row, 1954.

Bradshaw, John. *Dog Sense: How the New Science of Dog Behavior Can Make You a Better Friend to Your Pet.* New York: Basic Books, 2011.

Braithwaite, Victoria. *Do Fish Feel Pain?* New York: Oxford University Press, 2010.

Brechbuhl, Julien, Magali Klaey, and Marie-Christine Broillet. "Grueneberg Ganglion Cells Mediate Alarm Pheromone Detection in Mice." *Science* 321 (2008): 1092–95.

Brown, Arthur. "Grief in the Chimpanzee." *American Naturalist* 13 (1879): 173–75.

Brown, Guy. *The Living End: The Future of Death, Aging and Immortality*. New York: Macmillan, 2008.

Buchanan, Brett. *Onto-Ethologies: The Animal Environments of Uexkull, Heidegger, Merleau-Ponty, and Deleuze*. Albany, New York: SUNY Press, 2008.

Burns, John F. "The Vagabond Cat That Came to Stay." *New York Times*, July 25, 2010. Week in Review, 1.

Byock, Ira. *Dying Well: Peace and Possibilities at the End of Life*. New York: Riverhead Trade, 1998.

———. "Rediscovering Community at the Core of the Human Condition and Social Covenant." In "Access to Hospice Care: Expanding Boundaries, Overcoming Barriers." Supplement to *Hastings Center Report* 33, no. 2 (March–April 2003): 40–44.

Calarco, Matthew. "Thinking through Animals: Reflections on the Ethical and Political Stakes of the Question of the Animal in Derrida." *Oxford Literary Review* 29 (2007): 1–16. doi 10.3366/E0305149807000053.

———. *Zoographies: The Question of the Animal from Heidegger to Derrida*. New York: Columbia University Press, 2008.

Carbone, Larry. *What Animals Want: Expertise and Advocacy in Laboratory Animal Welfare Policy*. New York: Oxford University Press, 2004.

Carter, Zoe Fitzgerald. *Imperfect Endings: A Daughter's Tale of Life and Death*. New York: Simon & Schuster, 2010.

Chevalier-Skolnikoff, Suzanne, and Jo Liska. "Tool Use by Wild and Captive Elephants." *Animal Behaviour* 46 (1993): 209–19.

Coetzee, J. M. *Disgrace*. New York: Viking, 1999.

———. *The Lives of Animals*. Princeton, NJ: Princeton University Press, 2001.

Cooney, Kathleen. *In-Home Pet Euthanasia Techniques*. Loveland, CO: Home to Heaven, 2011. E-book.

Cooper, Gwen. *Homer's Odyssey*. New York: Delacorte Press, 2009.

Cronquist, Mark. "Livestock Mortality Management." Small Acreage Series, Colorado State University Cooperative Extension,, 2007.

Cryonics Institute. "Cryonics: A Basic Introduction." 2010. http://www .cryonics.org/prod.html.

Curlin, Farr A., Chinyere Nwodim, Jennifer L. Vance, Marshall H. Chin, and John D. Lantos. "To Die, to Sleep: US Physicians' Religious and Other Objections to Physician-Assisted Suicide, Terminal Sedation, and Withdrawal of Life Support," *American Journal of Hospice and Palliative Care* 25(2008): 112–20.

Dagg, Anne Innis. *The Social Behavior of Older Animals*. Baltimore: Johns Hopkins University Press, 2009.

Dawkins, Marian Stamp. "The Science of Suffering." In *Mental Health and Well-Being in Animals*, edited by Franklin D. McMillan, 47–56. Oxford: Blackwell, 2005.

———. "The Scientific Basis for Assessing Suffering in Animals." In *In Defense of Animals*, edited by Peter Singer, 27–40. New York: Basil Blackwell, 1985.

DeLennart, Eleonara. *Dogs Don't Cry Tears*. New York: Big Apple Vision Books, 2005.

Derrida, Jacques. *The Animal That Therefore I Am*. Translated by David Wills. New York: Fordham University Press, 2008.

———. *Aporias*. Translated by Thomas Dutoit. Stanford, CA: Stanford University Press, 1993.

DeStefano, Stephen. *Coyote at the Kitchen Door: Living with Wildlife in Suburbia*. Cambridge, MA: Harvard University Press, 2010.

De Waal, Frans, and Frans Lanting. *Bonobo: The Forgotten Ape*. Berkeley: University of California Press, 1997.

Dickey, James. "The Heaven of Animals." In *The Whole Motion: Collected Poems, 1945–1992*. Middletown, CT: Wesleyan University Press, 1992.

Dickinson, Janis L., and Miyoko Chu. "Animal Funerals: Do Magpies Express Grief?" *BirdScope*, vol. 21, no. 1 (2007). http://www.birds.cornell.edu/Publications/Birdscope/Winter2007/animal_funerals.html.

Didion, Joan. *The Year of Magical Thinking*. New York: Random House, 2005.

Dodman, Nicholas. *Good Old Dog*. New York: Houghton Mifflin Harcourt, 2010.

Dosa, David M. "A Day in the Life of Oscar the Cat." *New England Journal of Medicine* 357 (2007): 328–29.

———. *Making Rounds with Oscar*. New York: Hyperion, 2010.

Doty, Mark. *Dog Years: A Memoir*. New York: Harper Collins, 2007.

Douglas-Hamilton, Iain, Shivani Bhalla, George Wittemyer, and Fritz Vollrath. "Behavioural Reactions of Elephants towards a Dying and Deceased Matriarch." *Applied Animal Behaviour Science* 100 (2006): 87–102.

Downing, Robin. "Pain Management for Veterinary Palliative Care and Hospice Patients." In *Palliative Medicine and Hospice Care*, edited by Tami Shearer, 531–50. Veterinary Clinics of North America: Small Animal Practice, vol. 41, no. 3. Philadelphia: W. B. Saunders Company, 2011.

Dussel, Veronica, Steven Joffe, Joanne M. Hilden, Jan Watterson-Schaeffer, Jane C. Weeks, and Joanne Wolfe. "Considerations about Hastening Death among Parents of Children Who Die of Cancer." *Archives of Pediatrics and Adolescent Medicine* 164 (2010):231–37.

Dye, Dan, and Mark Beckloff. *Amazing Gracie.* New York: Workman, 2003.

Eberle, Scott. *The Final Crossing: Learning to Die in Order to Live.* Big Pine, CA: Lost Borders Press, 2006.

Ellis, Coleen. *Pet Parents: A Journey through Unconditional Love and Grief.* Bloomington, IN: IUniverse, 2011.

Engh, Anne L, Jacinta C. Beehner, Thore J. Bergman, Patricia L. Whitten, Rebekah R. Hoffmeier, Robert M. Seyfarth, and Dorothy L. Cheney. "Behavioral and Hormonal Responses to Predation in Female Chacma Baboons." *Proceedings of the Royal Society B* 273 (2006):707–12. http://rspb.royalsocietypublishing.org/content/273/1587/707.full.pdf+html.

Fish, Richard E., Marilyn J. Brown, Peggy J. Danneman, and Alicia Z. Karas. *Anesthesia and Analgesia in Laboratory Animals.* 2d ed. Amsterdam: Elsevier, 2008.

Flannery, Tim. "Getting to Know Them." *New York Review of Books,* April 29, 2011, 12–16.

Foer, Jonathan Safran. *Eating Animals.* New York: Little, Brown, 2009.

Fogle, Bruce. *The Dog's Mind: Understanding Your Dog's Behaviour.* Hoboken, NJ: Wiley, 1990.

Fox, Michael W. *Dog Body, Dog Mind: Exploring Canine Consciousness and Total Well-Being.* Guilford, CT: Lyons Press, 2007.

Franklin, Jon. *The Wolf in the Parlor.* New York: St. Martin's Griffin, 2009.

Fudge, Erica. *Pets.* Durham, UK: Acumen, 2008.

Fuller, Todd K., L. David Mech, and Jean Fitts Cochrane. "Wolf Population Dynamics." In *Wolves: Behavior, Ecology, and Conservation,* edited by David L. Mech and Luigi Boitani, 161–91. Chicago: University of Chicago Press, 2003.

Gawande, Atul. "Letting Go: What Should Medicine Do When It Can't Save Your Life?" *New Yorker,* August 2, 2010. http://www.newyorker.com/reporting/2010/08/02/100802fa_fact_gawande.

Gilbert, Sandra M. *Death's Door: Modern Dying and the Ways We Grieve.* New York: W. W. Norton, 2006.

Goetsch, Douglas. "Different Dogs." *New Yorker,* January 17, 2011, 60.

Goldenberg, Jamie L., Tom Pyszczynski, Jeff Greenberg, Sheldon Solomon, Benjamin Kluck, and Robin Cornwell. "I Am *Not* an Animal: Mortality

Salience, Disgust, and the Denial of Human Creatureliness." *Journal of Experimental Psychology* 130 (2001): 427–35.

Goodall, Jane. *Through a Window: My Thirty Years with the Chimpanzees of Gombe*. Boston: Houghton Mifflin, 1990.

Gopnik, Adam. "Dog Story: How Did the Dog Become Our Master?" *New Yorker*, August 8, 2011. http://www.newyorker.com/reporting/2011/08/08/110808fa_fact_gopnik.

Greenberg, J., S. Solomon, and T. Pyszcznski. "Terror Management Theory of Self-Esteem and Cultural Worldviews: Empirical Assessments and Conceptual Refinements." *Advances in Experimental Social Psychology* 29 (1997): 61–142.

Grenier, Roger. *The Difficulty of Being a Dog*. Translated by Alice Kaplan. Chicago: University of Chicago Press, 2000.

Grey Muzzle Organization. "Old Dogs and Animal Shelters." 2008. http://www.greymuzzle.org/PDF/OldDogsandAnimalShelters.aspx

Griffin, Donald R. *The Question of Animal Awareness*. New York: Rockefeller University Press, 1981.

Gustafsson, Lars. "Elegy for a Dead Labrador" and "The Dog." In *The Stillness of the World before Bach*. Edited by Christopher Middleton. Translated by Robin Fulton, Phillip Martin, Yvonne L. Sandstroem, Harriett Watts, and Christopher Middleton. New York: New Directions Books, 1988.

Hains, Bryan C. *Pain*. New York: Infobase, 2007.

Harris, Julia. *Pet Loss: A Spiritual Guide*. New York: Lantern Books, 2002.

Hatkoff, Amy. *The Inner World of Farm Animals*. New York: Steward, Tabori & Chang, 2009.

Hearne, Vicki. *Animal Happiness: A Moving Exploration of Animals and Their Emotions*. New York: Skyhorse, 2007.

Herzog, Hal. *Some We Love, Some We Hate, Some We Eat: Why It's So Hard to Think Straight about Animals*. New York: Harper, 2010.

Holekamp, Kay E., and Laura Smale. "Behavioral Development in the Spotted Hyena." *BioScience* 48 (1998): 997–1005.

Horowitz, Alexandra. *Inside of a Dog*. New York: Scribner, 2009.

Hoskins, Johnny. *Geriatrics and Gerontology of the Dog and Cat*. Philadelphia: Saunders, 2003.

Howell, Phillip. "A Place for the Animal Dead: Pets, Pet Cemeteries and Animal Ethics in Late Victorian Britain." *Ethics, Place and Environment* 5 (2002): 5–22.

Human Rights Watch. "Please, Do Not Make Us Suffer Any More . . ." : Access to Pain Treatment as a Human Right." 2009. http://www.hrw.org/en/reports/2009/03/02/please-do-not-make-us-suffer-any-more.

Ilardo, Joseph A. *As Parents Age: A Psychological and Practical Guide.* Acton, MA: VanderWyk & Burnham, 1995.

Institute for Laboratory Animal Research. *Recognition and Alleviation of Pain and Distress in Laboratory Animals.* Washington, DC: National Academies Press, 1992.

International Association for Animal Hospice and Palliative Care. "Canine Grief: 'Do Dogs Mourn?'" 2010. http://iaahpc.org/index.php/for-pet-parents/helpful-articles/item/8-canine-grief-%E2%80%93-do-dogs-mourn.

International Association for Hospice and Palliative Care. *The IAHPC Manual of Palliative Care.* 2d ed. 2008. http://www.hospicecare.com/manual/toc-main.html.

International Association for the Study of Pain. "Do Animal Models Tell Us about Human Pain?" *Pain Clinical Updates* 18, no. 5 (2010): 1–5.

———. "IASP Taxonomy." 2011. http://www.iasp-pain.org/AM/Template.cfm?Section=Pain_Defi . . . isplay.cfm&ContentID=1728.

International Institute for Animal Law. "Humane Euthanasia of Animals." 2010. http://www.animallaw.com/Humaneeuthanasia.htm.

Jeffers, Robinson. "The House-Dog's Grave." In *The Selected Poetry of Robinson Jeffers.* Edited by Tim Hunt. Stanford, CA: Stanford University Press, 2002.

Jennings, Bruce, True Ryndes, Carol D'Onofrio, and Mary Ann Baily. "Access to Hospice Care: Expanding Boundaries, Overcoming Barriers." Supplement to *Hastings Center Report,* vol. 33, no. 2 (March–April 2003).

Johns, Bud, ed. *Old Dogs Remembered.* San Francisco: Synergistic Press, 1993.

Jones, Amanda C., and Samuel D. Gosling. "Temperament and Personality in Dogs (*Canis familiaris*): A Review and Evaluation of Past Research." *Applied Animal Behavior Science* 95 (2005): 1–53.

Kass, P H., John C. New Jr., Jennifer M. Scarlett, and M. D. Salman. "Understanding Animal Companion Surplus in the United States: Relinquishment of Nonadoptables to Animal Shelters for Euthanasia." *Journal of Applied Animal Welfare Science* 4, no. 4 (2001): 237–48.

Katz, Jon. *Going Home: Finding Peace When Pets Die.* New York: Random House, 2011.

Kaufman, Sharon R. *And a Time to Die: How American Hospitals Shape the End of Life.* Chicago: University of Chicago Press, 2005.

Kay, Nancy. *Speaking for Spot: Be the Advocate Your Dog Needs to Live a Happy, Healthy, Longer Life.* North Pomfret, VT: Trafalgar Square Books, 2008.

Kerasote, Ted. *Merle's Door: Lessons from a Freethinking Dog.* New York: Harcourt, 2007.

Kiernan, Stephen P. *Last Rights: Rescuing the End of Life from the Medical System.* New York: St. Martin's Press, 2006.

Kitchell, Ralph, and Michael Guinan. "The Nature of Pain in Animals." In *The Experimental Animal in Biomedical Research.* Edited by Bernard E. Rollin and M. Lynne Kesel, 1:185–205. Boca Raton, FL: CRC Press, 1990.

Klarsfeld, André, and Frédéric Revah. *The Biology of Death: Origins of Mortality.* Translated by Lydia Brady. Ithaca, NY: Cornell University Press, 2004.

Kluger, Jeffrey. "Inside the Minds of Animals." *Time Magazine,* August 16, 2010, 36–43.

Koktavy, Doug. *The Legacy of Beezer and Boomer: Lessons of Living and Dying from My Canine Brothers.* Denver: BBrothers Press, 2010.

Kreeger, Terry. "The Internal Wolf: Physiology, Pathology, and Pharmacology." In *Wolves: Behavior, Ecology, and Conservation.* Edited by David L. Mech and Luigi Boitani, 192–217. Chicago: University of Chicago Press, 2003.

Kübler-Ross, Elizabeth. *On Death and Dying.* New York: Scribner, 1997.

Kuzniar, Alice A. *Melancholia's Dog.* Chicago: University of Chicago Press, 2006.

Leigh, Diane, and Marilee Geyer. *One at a Time: A Week in an American Animal Shelter.* Santa Cruz, CA: No Voice Unheard, 2003.

Levine, Stephen. *A Year to Live.* New York: Three Rivers Press, 1997.

Lindsay, Stephen R. *Handbook of Applied Dog Behavior.* Vol. 1, *Adaptation and Learning.* Ames: Iowa State University Press, 2000.

Lindsey, Jennifer, and Jane Goodall. *Jane Goodall: 40 Years at Gombe.* New York: Stewart, Tabori & Chang, 1999.

Lorenz, Konrad. *Man Meets Dog.* New York: Penguin Books, 1954.

Losey, Robert J., Vladimir I. Bazaliiskii, Sandra Garvie-Lok, Mietje Germonpré, Jennifer A. Leonard, Andrew L. Allen, M. Anne Katzenberg, and Mikhail V. Sablin. "Canids as Persons: Early Neolithic Dog and Wolf Burials, Cis-Baikal, Siberia." *Journal of Anthropological Archaeology* 30, no. 2 (2011): 174–89.

MacNulty, Daniel R., Douglas W. Smith, John A. Vucetich, L. David Mech, Daniel R. Stahler, and Craig Packer. "Predatory Senescence in Ageing Wolves." *Ecology Letters* 12 (2009):1347–56.

Mahoney, James. *Saving Molly: A Research Veterinarian's Choices.* Chapel Hill, NC: Algonquin Books, 1998.

Mann, Thomas. *Bashan and I.* 1923. Reprint, Philadelphia: Pine Street Books, 2003.

Marino, Susan. *Getting Lucky.* New York: Stewart, Tabori & Chang, 2005.

McComb, Karen, Lucy Baker, and Cynthia Moss. "African Elephants Show High Levels of Interest in the Skulls and Ivory of Their Own Species." *Biology Letters* 2, no. (2006): 26–28.

McComb, Karen, Graeme Shannon, Sarah M. Durant, Katito Sayialel, Rob Slotow, Joyce Poole, and Cynthia Moss. "Leadership in Elephants: The Adaptive Value of Age." *Proceedings of the Royal Society* B 278 (2011): 3270–76.

McCullough, Susan. *Senior Dogs for Dummies.* Hoboken, NJ: Wiley, 2004.

McFarland, David, ed. *The Oxford Companion to Animal Behavior.* New York: Oxford University Press, 1982.

McFarlane, Rodger, and Philip Bashe. *The Complete Bedside Companion: No-Nonsense Advice on Caring for the Seriously Ill.* New York: Simon & Schuster, 1998.

McMillan, Franklin D. "The Concept of Quality of Life in Animals." In *Mental Health and Well-Being in Animals,* edited by Franklin D. McMillan, 191–200. Oxford: Blackwell, 2005.

———. "Do Animals Experience True Happiness?" In *Mental Health and Well-Being in Animals,* edited by Franklin D. McMillan, 221–33. Oxford: Blackwell, 2005.

———, ed. *Mental Health and Well-Being in Animals.* Oxford: Blackwell, 2005.

McMillan, Franklin D., with Kathryn Lance. *Unlocking the Animal Mind.* Emmaus, PA: Rodale, 2004.

Mech, L. David, and Luigi Boitani, eds. *Wolves: Behavior, Ecology, and Conservation.* Chicago: University of Chicago Press, 2003.

Mehta, Pranjal H., and Samuel D. Gosling. "Bridging Human and Animal Research: A Comparative Approach to Studies of Personality and Health." *Brain, Behavior, and Immunity* 22 (2008): 651–61.

Mendl, Michael, Julie Brooks, Christine Basse, Oliver Burman, Elizabeth Paul, Emily Blackwell, and Rachel Casey. "Dogs Showing Separation-Related Behavior Exhibit a 'Pessimistic' Cognitive Bias." *Current Biology* 20 (2010): R839–R840.

Mendl, Michael, Oliver H. P. Burman, and Elizabeth S. Paul. "An Integrative and Functional Framework for the Study of Animal Emotion and Mood." *Proceedings of the Royal Society* B 277 (2010): 2895–2904.

Miller, Lisa. *Heaven: Our Enduring Fascination with the Afterlife*. New York: Harper, 2010.

Miller, Sara Swan. *Three Stories You Can Read to Your Dog*. New York: Scholastic, 1995.

Morris, Desmond. *Dog Watching*. New York: Three Rivers Press, 1986.

Moss, Cynthia. *Elephant Memories*. Chicago: University of Chicago Press, 2000.

Nagasawa, Miho, and Mitsuaki Ohta. "The Influence of Dog Ownership in Childhood on the Sociality of Elderly Japanese Men." *Animal Science Journal* 81 (2010): 377–83.

Nakaya, Shannon Fujimoto. *Kindred Spirit, Kindred Care*. Novato, CA: New World Library, 2005.

National Academy of Sciences. "Recognizing Pain in Animals—Dogs." 2010. http://dels-old.nas.edu/animal_pain/dogs.shtml.

National Centre for the Replacement, Refinement, and Reduction of Animals in Research. "Euthanasia." 2011. http://www.nc3rs.org.uk/category .asp?catID=15

National Research Council, Committee on Pain and Distress in Laboratory Animals. *Recognition and Alleviation of Pain and Distress in Laboratory Animals*. Washington, DC: National Academy Press, 1992.

New, John C., Jr., William J. Kelch, Jennifer M. Hutchison, Mo D. Salman, Mike King, Janet M. Scarlett, and Phillip H. Kass. "Birth and Death Rate Estimates of Cats and Dogs in U.S. Households and Related Factors." *Journal of Applied Animal Welfare Science* 7 (2004): 229–41.

Odendaal, J. S., and R. A. Meintjes. "Neurophysiological Correlates of Affiliative Behavior between Humans and Dogs." *Veterinary Journal* 165 (2003): 296–301.

Osvath, Mathias. "Spontaneous Planning for Future Stone Throwing by Male Chimpanzee." *Current Biology* 19 (2009): R190–R191.

Pacelle, Wayne. *The Bond: Our Kinship with Animals, Our Call to Defend Them.* New York: HarperCollins, 2011.

Packard, Jane M. "Wolf Behavior: Reproductive, Social, and Intelligent." In *Wolves: Behavior, Ecology, and Conservation,* edited by David L. Mech and Luigi Boitani, 35–65. Chicago: University of Chicago Press, 2003.

Panksepp, Jaak. "Affective-Social Neuroscience Approaches to Understanding Core Emotional Feelings in Animals." In *Mental Health and Well-Being in Animals,* edited by Franklin D. McMillan, 57–75. Oxford: Blackwell, 2005.

Patronek, Gary J., and Andrew N. Rowan. "Determining Dog and Cat Numbers and Population Dynamics." *Anthrozoos* 8, no. 4 (1995): 199–205.

Patterson, Charles. *The Eternal Treblinka: Our Treatment of Animals and the Holocaust.* New York: Lantern Books, 2002.

Peeters, Marijke E., and Jolle Kirpensteijn. "Comparison of Surgical Variables and Short-Term Postoperative Complications in Healthy Dogs Undergoing Ovariohysterectomy or Ovariectomy." *Journal of the American Veterinary Medical Association* 238 (2011): 189–94.

Pinney, Chris C. *The Complete Home Veterinary Guide.* 3d ed. New York: McGraw-Hill, 2003.

———. *Vizslas: A Complete Pet Owner's Manual.* Hauppauge, NY: Barron's Educational Series, 1998.

Pomerance, Diane. *Animal Elders: Caring about Our Aging Animal Companions.* Flower Mound, TX: Polaire Publications, 2005.

Poole, Joyce. *Coming of Age with Elephants.* New York: Hyperion Books, 1996.

———. *Elephants.* Stillwater, MN: Voyageur Press, 1997.

Raby, C. R., D. M. Alexis, A. Dickinson, and N. S. Clayton. "Planning for the Future by Western Scrub-Jays (*Aphelocoma Californica*): Implications for Social Cognition." *Animal Behaviour* 70 (2007): 1251–63.

Ramsden, Edmund and Duncan Wilson. "The Nature of Suicide: Science and the Self-Destructive Animal." *Endeavour* 34 (2010): 21–24. doi: 10.1016/j.endeavour.2010.01.005.

Randour, Mary Lou. *Animal Grace: Entering a Spiritual Relationship with Our Fellow Creatures.* Novato, CA: New World Library, 2000.

Rawls, Wilson. *Where the Red Fern Grows.* New York: Yearling, 1996.

Reck, Julie. *Facing Farewell: A Guide for Making End of Life Decisions for Your Pet.* N.p.: Lulu, 2010.

Reeve, Charlie L., Steven G. Rogelberg, Christiane Spitzmüller, Natalie

DiGiacomo, et al. "The Caring-Killing Paradox: Euthanasia-Related Strain among Animal-Shelter Workers." *Journal of Applied Social Psychology* 35 (2005): 119–43.

Reitman, Judith. "From the Leash to the Laboratory." *Atlantic Monthly* 286, no. 1 2000): 17–21. http://www.theatlantic.com/past/docs/issues/2000/07/reitman.htm.

Rilke, Rainer Maria. *Duino Elegies.* Translated by Dora Van Franken and Roger Nicholson Pierce. Longmont, CO: Center of Balance Press, 2011.

Rivera, Michelle. *Hospice Hounds: Animals and Healing at the Borders of Death.* New York: Lantern Books, 2001.

———. *On Dogs and Dying: Inspirational Stories of Hospice Hounds.* West Lafayette, IN: Purdue University Press, 2010.

Roberts, William A., Miranda C. Feeney, Krista MacPherson, Mark Petter, Neil McMillan, and Evanya Musolina. "Episodic-Like Memory in Rats: Is It Based on When or How Long Ago?" *Science* 320 (2008): 113–15. doi: 10.1126/science.1152709.

Rollin, Bernard E. *An Introduction to Veterinary Medical Ethics: Theory and Cases.* 2d ed. Oxford: Blackwell, 2006.

———. *Science and Ethics.* New York: Cambridge University Press, 2006.

———. *The Unheeded Cry: Animal Consciousness, Animal Pain, and Science.* Expanded ed. Ames: Iowa State University Press, 1989.

Rozin, P., J. Haidt, and C. McCauley. "Disgust." In *Handbook of Emotions,* edited by M. Lewis and J. Haviland, 637–53. New York: Guilford Press, 1993.

Rylant, Cynthia. *Dog Heaven.* New York: Blue Sky Press, 1995.

Sakson, Sharon. *Paws and Effect: The Healing Power of Dogs.* New York: Spiegel & Grau, 2009.

Sanders, Clinton R. "Killing with Kindness: Veterinary Euthanasia and the Social Construction of Personhood." *Sociological Forum* 10 (1995): 195–214.

San Francisco Society for the Prevention of Cruelty to Animals. *Fospice Program Manual.* 2011. http://www.sfspca.org/sites/default/files/Fospice_Manual.pdf.

Schaffer, Michael. *One Nation under Dog.* New York: Henry Holt, 2009.

Schoen, Allen M. *Kindred Spirits: How the Remarkable Bond between Humans and Animals Can Change the Way We Live.* New York: Broadway Books, 2001.

Schoen, Allen M., and Pam Proctor. *Love, Miracles, and Animal Healing*. New York: Simon & Schuster, 1995.

Schultz, Krista. "An Emerging Occupational Threat? Study Seeks Reasons for High Suicide Rate among Veterinarians" *DVM Newsmagazine*. 2008. http://veterinarynews.dvm360.com/dvm/Veterinary+business/ An-emerging-occupational-threat/ArticleStandard/Article/detail/ 514794.

Schwartz, Marion. *A History of Dogs in the Early Americas*. New Haven, CT: Yale University Press, 1997.

Seeger, Ruth Crawford. *Animal Folk Songs for Children*. Garden City, NY: Doubleday, 1950.

Serpell, James, ed. *The Domestic Dog: Its Evolution, Behaviour and Interactions with People*. New York: Cambridge University Press, 1995.

———. *In the Company of Animals: A Study of Human-Animal Relationships*. New York: Basil Blackwell, 1986.

Shearer, Tami, ed. *Palliative Medicine and Hospice Care*. Vol. 41, no. 3 of Veterinary Clinics of North America: Small Animal Practice. Philadelphia: W. B. Saunders, 2011.

Shriver, Adam. "Knocking out Pain in Livestock: Can Technology Succeed Where Morality Has Stalled?" *Neuroethics* 2, no. 3 (2009): 115–24. doi: 10.1007/s12152–009-9048–6.

Siebert, Charles. "New Tricks." *New York Times Magazine*, April 8, 2007. http:// www.nytimes.com/2007/04/08/magazine/08animal.t.html?_r=1.

Sife, Wallace. *The Loss of a Pet*. 3d ed. Hoboken, NJ: Wiley, 2005.

Stafford, Kevin. *The Welfare of Dogs*. Dordrecht: Springer, 2006.

Stein, Garth. *The Art of Racing in the Rain: A Novel*. New York: Harper, 2009.

Steward, Kelly. *Gorillas: Natural History and Conservation*. Stillwater, MN: Voyageur Press, 2003.

Stoddard, Sandol. *The Hospice Movement: A Better Way of Caring for the Dying*. New York: Vintage, 1992.

Tannenbaum, Jerrold. *Veterinary Ethics: Animal Welfare, Client Relations, Competition and Collegiality*. 2d ed. St. Louis, MO: Mosby, 1995.

Task Force on Palliative Care. *Last Acts: Precepts of Palliative Care*. 1997. http://www.aacn.org/WD/Palliative/Docs/2001Precep.pdf.

Taylor, David. *Old Dog, New Tricks: Understanding and Retraining Older and Rescued Dogs*. Buffalo, NY: Firefly Books, 2006.

Villalobos, Alice. "Bringing Pawspice to Your Practice." *Veterinary Practice News.* 2009. http://www.veterinarypracticenews.com/vet-practice-news-columns/bond-beyond/pawspice.aspx.

———. "Quality of Life Scale." *Veterinary Practice News.* 2006. http://www.veterinarypracticenews.com/vet-practice-news-columns/bond-beyond/quality-of-life-scale.aspx.

von Uexküll, Jacob. "A Stroll through the Worlds of Animals and Men." In *Instinctive Behavior: The Development of a Modern Concept,* edited by C. H. Schiller, 5–80. 1934. Reprint, New York: International Universities Press, 1957.

Wall, Patrick. *Pain: The Science of Suffering.* New York: Columbia University Press, 2000.

Watkins, Tom. "Paper Delves into British Veterinarians' High Suicide Risk." CNN *World.* 2010. http://current.com/http://www.cnn.com/2010/WORLD/europe/03/26/england.veterinarians.suicide/index.html?hpt=T2.

Watts, Heather E., and Kay E. Holekamp. "Ecological Determinants of Survival and Reproduction in the Spotted Hyena." *Journal of Mammology* 90 (2009): 461–71.

Weiner, Jonathan. *Long for This World: The Strange Science of Immortality.* New York: Ecco/HarperCollins, 2010.

Weingarten, Gene, and Michael S. Williamson. *Old Dogs Are the Best Dogs.* New York: Simon & Schuster, 2008.

Whitehead Hal. *Sperm Whales: Social Evolution in the Ocean.* Chicago: University of Chicago Press, 2003.

Williams, J. M., E. V. Lonsdorf, M. L. Wilson, J. Schumacher-Stankey, J. Goodall, and A. E. Pusey. "Causes of Death in the Kasekela Chimpanzees of Gombe National Park, Tanzania." *American Journal of Primatology* 70(2008): 766–77.

Winograd, Nathan J. *Redemption: The Myth of Pet Overpopulation and the No Kill Revolution in America.* Santa Clara, CA: Almaden Press, 2009.

Wolfelt, Alan D. *When Your Pet Dies: A Guide to Mourning, Remembering, and Healing.* Fort Collins, CO: Companion Press, 2004.

Woolf, Virginia. *Flush.* New York: Harcourt Brace Jovanovich, 1933.

Yao, M., J. Rosenfeld, S. Attridge, S. Sidhu, V. Aksenov, and C. D. Rollo. "The Ancient Chemistry of Avoiding Risks of Predation and Disease." *Evolutionary Biology* 36 (2009):267–81.

Yeats, William Butler. "Death." In *The Collected Poems of W. B. Yeats*. Edited by Richard Finneran, 234. 2d rev. ed., New York: Scribner, 1996.

Zhou, Wenyi, and Jonathan D. Crystal. "Evidence for Remembering When Events Occurred in a Rodent Model of Episodic Memory." *Proceedings of the National Academy of Sciences of the United States of America* 106 (2009): 9525–29.

Index